The Singularity of Hope

Humanity's Role in an AI-Dominated Future

Sam Sammane

Disclaimer

Table of Contents

Dedication

To my father, **Ratib**,

Though you never saw me reach this place, your spirit has been my guiding light. Your unwavering belief in me, your enduring values, and the memories we shared have been the silent companions in my journey of writing this book. This work stands as a testament to your enduring influence on my life.

You may not be here to witness this moment, but your legacy lives on in these pages.

With love and remembrance,

Sam Sammane

Acknowledgments

As I sit down to express my deepest appreciation, I'm reminded of the incredible journey that led to these pages. This book, a tapestry woven from many hands and hearts, wouldn't have been possible without the love, support, and wisdom of so many.

To my beloved family, the unshakable foundation of my life: Words can scarcely capture the depth of my gratitude. To my wife, Nour, my guiding light and partner, your support and love have been the lifeblood of my aspirations. To my daughters, Zena, Sarah, and Ayah, your joy and laughter have been my sanctuary amidst the whirlwind of creativity. To my sister Reem, whose wisdom and encouragement have always guided me, to my brother Waseem, a constant source of strength and camaraderie, and to my mother Souad, the epitome of resilience and love, your collective belief in my dreams, your endless patience through the long hours, and your unconditional love have been the very pillars of my strength and perseverance.

To my extended family, though unmentioned by name, your influence is deeply felt and cherished. To each and every one of you, thank you from the bottom of my heart. You are my home and my haven.

To my mentor, Dominique Borrione, a guiding star in my pursuit of knowledge: Your profound insights and mentorship have not only shaped this book but have also deepened my understanding of the intricate world of technology and AI. In times of doubt, your wisdom has been my lighthouse.

A heartfelt thanks to my best friends Samir Mounir and Muhammad Alsayid. Your candid feedback and our countless late-night discussions have added layers of depth to this work.

Immense gratitude to the team at Authors On Mission, Jonathan, Asif, Lanie, Lindsay, and, more particularly, Vikrant Shaurya. Your expertise, patience, and dedication have polished these words into a narrative I am honored to share.

I extend my thanks to the many experts and professionals in AI, theorem proving, and technology. In particular, J Strother Moore, Matt Kaufmann, and Warren A. Hunt, Jr. Your willingness to share your knowledge and perspectives has greatly enriched me as a person and an author.

To Hesham Zreik, Chadi Bitar, Gabriel Kafi, Dia Abochamh, Abgar Barseyten, Jochen Zoller and all my friends, near and far, who have journeyed alongside me – your encouragement, friendship, and belief in my vision have been sources of unwavering support.

And thank you, dear reader, for stepping into this journey with me. Your curiosity and engagement are what truly bring this book to life.

This book is not just a product of my efforts but a testament to the collective support and belief of each one of you. Thank you for being an integral part of this remarkable journey.

With gratitude,

Sam Sammane

Preface

In the nascent stages of my foray into the realms of artificial intelligence and its intricate dance with humanity, I was both captivated and troubled. The swift progression of AI technology, the fervent debates over its ethical ramifications, and its profound influence on the future fabric of human society didn't just pique my academic curiosity—they were the drumbeats to a deeper, more profound understanding. In this crucible of wonder and worry, "The Singularity of Hope" took shape.

This book isn't an academic discourse on artificial intelligence. It's a narrative, a heartfelt journey into the essence of our humanity in an era increasingly orchestrated by machines. My mission is to unravel the complexities of AI and shed light on the subtleties of this technological phenomenon that is a product and a progenitor, a tool and a sovereign.

Penning "The Singularity of Hope" became a personal odyssey that paralleled my life's journey from a young, inquisitive mind to an innovator, standing at the intersection of technology and human experience. Each chapter and revelation resonated with my own experiences and reflections, from the early days of tinkering in my backyard to the profound realizations about technology's role in shaping our destiny.

Driven by a singular, burning question—What role does humanity play in an AI-dominated future?—I delved deep. And the more I explored, the clearer it became: the true power lies not within the circuits and codes of the machine but within us—our values, choices, and vision for what lies ahead.

"The Singularity of Hope" is more than a book; it's an invitation. An invitation to you, the reader, particularly those who will inherit and shape this new world, to ponder the profound questions posed by AI. It's a plea to look past the silicon and the algorithms to grapple with the moral and ethical crossroads we face. It's a call to imagine a future where technology doesn't overshadow our humanity but amplifies it.

As you turn these pages, I invite you to journey with me. Whether you're a student, an innovator, a policymaker, or merely someone who marvels at the world's wonders, I hope "The Singularity of Hope" enlightens, challenges, and, above all, sparks a conversation about our collective future in this age of AI—a conversation that I have lived, breathed, and loved in every step of my own life's path.

Introduction

When AI started correcting human grammar, we didn't see it as an act of supremacy but as an advanced spellchecker. No one got scared that it may some day take over the world. It was a subtle beginning, almost sympathetic. But as these spellcheckers grew, they began reshaping our world in ways we never anticipated. Gradually, what started as a mere spellchecker and digital assistant evolved with generative AI and matured into a giant spellchecker casting an illusion of human intelligence. These AI systems, sophisticated and intricate, began crafting entire articles, designing complex solutions, and even simulating human conversations with startling accuracy. Yet, beneath this veneer of intellect lies a profound question: is this genuine intelligence or merely a reflection of our own thoughts, organized in the giant data structure, masterfully echoed back to us?

The great irony of generative AI isn't that it has become too human-like but that it has revealed how machine-like humans have been in the last 50 years. What if, instead of viewing such AI systems as competitors against humanity in a race for superior intelligence, we simply saw them as tools, "super tools" even? As super tools, these AI systems don't replace but rather enlighten, offering us insights while valuing our human perspective in shaping a sustainable, just world. In this utopia, AI doesn't replace our work; it elevates it. We move beyond the routine, exploring arts, sciences, and the mysteries of our universe, our physical and intellectual capacities augmented by machines. This is a world where AI doesn't eclipse humanity but amplifies the very traits that make us human–our creativity, empathy, love, and the relentless pursuit of knowledge and wisdom.

The Singularity of Hope

Envision a future where the pinnacle of artificial intelligence isn't just a marvel of technology but a testament to human ingenuity and control. "The Singularity of Hope" is not merely a concept; it's a manifesto for a future where human intelligence doesn't surrender its reigns to AI but guides and molds it. In this book, we invite you on an extraordinary journey, one where we redefine the coexistence of humans and machines.

Picture this– a world where AI's immense capabilities are harnessed, directed, and amplified by human wisdom and insight. This is a future where AI becomes an extension of our will, not a replacement for it. A world where the symbiosis of human intellect and computational power solves our greatest challenges, elevates our creativity, and enriches our lives.

The core of 'The Singularity of Hope' is about empowering human intelligence to remain the architect of our destiny, even in the face of advanced AI. It's about creating AI that not only understands our instructions but also our intentions, our ethics, and our aspirations. This isn't about building machines that think for us; it's about developing AI that thinks with us, under our guidance and for our collective benefit.

This book is a call to action to technologists, thinkers, policymakers, and every individual who dreams of a future where technology is wielded with wisdom and foresight. It's a blueprint for an era where we don't just coexist with AI but command it, ensuring that it adheres to the values and principles that define us as humans.

If you've ever wondered about the role of human intelligence in an AI-driven future, "The Singularity of Hope" is your guide. It's an exploration and a roadmap for harnessing the power of AI while keeping the essence of humanity– our intellect, morality, and creative spirit–firmly at the helm.

Join us in this narrative, not just to understand AI but to shape it. This is about charting a course where human intelligence is not overshadowed but showcased, steering the ship of technological advancement towards a horizon of hope, dignity, and human excellence. "The Singularity of Hope" is more than a book; it's a declaration of our potential to guide, control, and excel in an AI-augmented world.

The Purpose of the Book

This is not another AI book; it's a message of hope, an algorithm that no machine could truly master, yet every human could understand. It is a call for humanity to embrace our intelligence and progress towards a shared vision for a prosperous world, hand in hand with their AI robots. A future where AI becomes humanity's greatest asset. Here, AI is not a threat but a bridge - a connection that deepens our understanding of artificial and human intelligence.

At its heart, this book is about empowering human intelligence. It's about ensuring that we don't lose sight of what makes us truly human in our quest to advance technology. We are the artists, the thinkers, the dreamers, and our intelligence will guide and shape the future of AI.

But it's not just about control; it's about collaboration. Imagine a world where AI doesn't replace us but enhances us. In a world where AI helps us solve the problems we can't solve alone, helps us understand the mysteries of the universe, and makes our lives better, richer, and more fulfilling.

However, with great power comes great responsibility. "The Singularity of Hope" is a call to engage in the ethical and moral implications of AI. It's about developing AI that aligns with our deepest values and serves the greater good of humanity.

We are calling on technologists, policymakers, and leaders to join us in shaping this future. It's a future that requires us to be educated, informed, and prepared for the changes AI will bring. It's about adaptability, lifelong learning, and staying ahead of the curve.

But let's be clear – we are not there yet. AI today is not the AI of science fiction. It has its limitations, and we must understand these. We need to bridge the gap between fantasy and reality, and that's what "The Singularity of Hope" aims to do.

Overview

Throughout the pages of "The Singularity of Hope", each chapter of this book is not just a chapter; it's a window into a future unfolding right before our eyes.

In Chapter 1, we embark on this journey, delving into the intricate landscape of human intelligence. It's a world of emotional, social, and experiential intelligence– realms where humans excel. Here, we contrast these uniquely human traits with AI's prowess in logical-mathematical and linguistic domains, exploring the current gaps in AI's ability to replicate the full spectrum of human cognition.
In Chapter 2, we turn to the rise of digital workers. We're entering an era where AI bots and robots are not just concepts but realities transforming our workplaces. This chapter isn't just about technological advancement;

it's about its impact on our economy, our jobs, and how we need to rethink our approach to work in an AI-dominated world.

Chapter 3 introduces a visionary concept: Human-AI-Augmentation (HAIA). This isn't about machines taking over; it's about them working in concert with us. It's a symphony of human and AI intelligence, creating harmony across sectors like healthcare, education, and the arts.

Then, in Chapter 4, we explore the edge of tomorrow– the AI singularity. It's a point where AI might start upgrading itself, evolving beyond our current understanding. But this chapter poses a critical question: Could this leap in AI's evolution be a collaborative effort with humanity rather than an independent stride?

Chapter 5 takes a deep, empathetic look at AI's impact on mental health. It's about understanding how AI in the workplace affects our well-being, our sense of purpose, and it urges organizations to take proactive steps to address these challenges.

In Chapter 6, we confront the "wealth paradox." This is where we see AI's double-edged sword– increasing affluence and widening inequality. We dive into the drivers of this paradox, like job automation and data monopolies, and propose innovative solutions such as Universal Basic Income and "AI investor" concepts for a more equitable future.

Finally, Chapter 7 takes us to a critical evaluation of AI surpassing human intelligence. It's a cautionary tale, warning of the risks of unregulated AI, but also a hopeful vision. Here, we imagine a "Singularity of Hope," a future where humans and AI don't just coexist but synergize, enriching humanity in ways we've only dreamed of.
Each chapter of this book isn't just a narrative; it's a stepping stone to a future that we can shape, a future where technology is harnessed for the greater good of humanity.

Then, we arrive at the Conclusion, a chapter that is not just a summary but a call to action for humanity. This isn't the end of the book; it's the beginning of a movement.

This is where we rally the collective intelligence, creativity, and spirit of humanity to steer the course of AI. It's a summons to each one of us – technologists, policymakers, educators, and citizens – to play an active role in shaping a future where AI and human intelligence don't just coexist but synergize to unlock unprecedented possibilities. This chapter reiterates our core theme of hope and determination. It's a blueprint for responsible innovation, a guide for ethical stewardship, and a manifesto for a future where technology serves to amplify our humanity, not diminish it. As we close this chapter, we open a new chapter in human history–one where we embrace the potential of AI with wisdom, foresight, and an unwavering commitment to the betterment of all.

Chapter 1

The Mind and The Machine

The field of AI has made incredible advances, but we are still very far from building systems that possess general intelligence, even at the level of a small child.

John McCarthy

Computer Scientist, known as one of the Founders of AI

The Mind Revealed

The Sorcerer of Software Intelligence

Once upon a time, in 1998, a young boy named Aaron had a gift for computers. One day, he wrote a special program where he could ask questions, and the computer would answer. It was like saying 'hello' to a friend, but this friend was a computer.

Aaron was so proud of what he had made that he couldn't wait to tell his grandma, Edith. With shining eyes, he told her, 'Grandma, I think my computer is alive!' Edith, who knew little about computers, was amazed. She looked at Aaron with pride and wondered, maybe, just maybe, that computer really was alive.

Fast forward to 2022, and Aaron was no longer a little boy but a data scientist. He was working on something big – AI systems- using a large language model. It was like his first 'hello world' program, but a million times more advanced.

This time, he wasn't just impressing his grandma but making the whole world wonder. With each question the AI answered, it seemed more and more like it had a mind of its own. Aaron's journey was about more than just writing programs that fool us. It was about dreams, wonders, and the magic of what we can make with technology.

Imagine a universe where the pinnacle of intelligence is the human brain, a realm where the mere idea of robots running offices and industries seems like a scene from a sci-fi movie. This is the world where Aaron's story unfolds – a tale not just about a boy who loved computers but a testament to the inspiring and creative capacity of technology.

Aaron's journey began with a young, curious mind that made his grandmother, Edith, ponder the possibility of a living computer. From those early days of innocent wonder, Aaron grew to become a man whose work with AI captivated the world. His path reflects the boundless potential of technology and imagination, challenging the dramatic tales of sci-fi movies where robots like the Terminator – devoid of feelings and unstoppable –shape our perceptions of AI.

Yet, alongside these tales of high-stakes action, there's a lighter side, reminiscent of shows like The Jetsons, where robots are friendly companions that make life easier. These contrasting views of technology – from fears of a robot uprising to miracles of AI integration – are understandable, given how intertwined AI has become with our lives and popular culture.

But let's get real: today's AI, the kind Aaron grew to master, isn't about mimicking the human mind or plotting world domination. It's about the elegance of algorithms, the sophistication of data analysis, and the seamless integration into our daily lives. Think of the AI that reroutes you on Google Maps, the phone camera that captures the perfect selfie, or the digital assistants that manage our schedules. Aaron's journey from his first 'hello world' program to developing advanced AI systems is a testament to the power of dreams and the endless possibilities in the art of programming.

From his first interaction with a computer screen to creating AI systems that are far more integrated and fascinating than any Hollywood narrative, Aaron's story mirrors our evolving relationship with technology – a blend of curiosity, innovation, and the ever-expanding horizons of the digital world.

The Intelligence of the Mind

Let's explore the multifaceted nature of human intelligence, as exemplified through Aaron's story. This examination delves into the various aspects of intelligence beyond mere academic knowledge or technical skill.

This multifaceted nature encompasses cognitive intelligence, which involves critical thinking and problem-solving; emotional intelligence, which highlights the management of emotions and impulses; social intelligence, which emphasizes interpersonal skills and communication; and experiential intelligence, which draws upon personal experiences and emotional depth.

In Aaron's adventures with computers, we see a beautiful display of various kinds of intelligence at play. It's not just about having book smarts or a lot of knowledge. Aaron's journey exemplifies cognitive intelligence, which is about thinking, decision-making, and understanding complex ideas. Every time he writes a new program, he's not just typing code; he's engaging in critical thinking and problem-solving, hallmarks of cognitive intelligence.

Emotional intelligence is another key element that Aaron demonstrates. Intelligence isn't just about logical thinking; it's also about how we handle our emotions and instincts. Aaron doesn't react impulsively when he encounters a difficult problem in his programming. Instead, he takes a step back, assesses the situation, and approaches it calmly and collectedly. This ability to manage emotions and impulses is crucial to emotional intelligence.

Aaron's story also highlights social intelligence. Human intelligence is special because it involves not only how we think but also how we relate to others. When Aaron excitedly shares his achievements with his grandmother, he's using his social intelligence. He understands and appreciates social bonds and communication, important parts of our intelligence.

Lastly, Aaron's experiences illustrate experiential intelligence. As he grows and achieves more in AI, his life experiences continue to shape his intelligence. The joy he feels when sharing his 'computer friend' with his grandmother and his later professional accomplishments all add to his experiential intelligence. His journey shows that our feelings, experiences, and the lessons we learn from them are integral to our overall intelligence.

Moving beyond Aaron's story, researchers have identified a plethora of intelligence types that define and enrich the human experience. While IQ tests often measure logical and mathematical intelligence, a vast world of other intelligences shape our abilities and talents. For example:

- Spatial Intelligence is like a hidden superpower in architects, engineers, and artists. It's the ability to visualize and manipulate objects in space, to see the world in dimensions and shapes, and to bring visions to life.
- Musical Intelligence resonates in the souls of songwriters, composers, and singers. It's the unique ability to understand and create music, to discern pitch, rhythm, and tone, and to transform notes into emotions and stories.
- Creative Intelligence: The ability to innovate, think abstractly, and create new ideas or art. This intelligence is crucial for artists, inventors, and anyone who thinks outside the box.
- Existential Intelligence is the ability to contemplate life's biggest questions. Philosophers and spiritual leaders, who ponder the meaning of existence, life, and death, embody this intelligence and guide others in their search for deeper understanding.
- Intrapersonal Intelligence is the journey within. It's crucial for psychologists, spiritual leaders, and reflective writers who seek to understand their own emotions, motivations, and inner states of being as they navigate the complex waters of the human psyche.

Each of these unique and diverse intelligences contributes to the rich tapestry of human capabilities, offering a wider and deeper understanding of what it means to be intelligent beyond conventional measures.

Being Human: Intuition and Emotion as Our Super Power

Consider intuition, that mysterious inner guide, an unspoken wisdom that often leads us to truth without needing conscious reasoning. It's like an internal compass, guiding us through life's complexities with an almost magical sense of knowing. This intuitive power, unique to humans, highlights our ability to make decisions that resonate with our deepest selves, often transcending the limits of logical thinking.

Then there's empathy, the ability to "step into someone else's shoes," to feel what they feel. Empathy is the golden thread that binds human relationships, fostering understanding and compassion. It allows us to connect deeply with others, to share their joys and sorrows, and to build communities based on mutual respect and care. In a world where AI can simulate conversation but not genuine emotion, empathy stands as a testament to the unique emotional intelligence of humans.

Love, in all its forms, is another cornerstone of human intelligence. It's a force that inspires, motivates, and uplifts. Love drives us to achieve, to protect, and to nurture. It's the foundation of families, friendships, and partnerships and fuels our most profound and enduring connections. The capacity to love and be loved is a uniquely human trait, one that no machine can replicate or understand.

Ambition, the fire that fuels our dreams and aspirations, is another uniquely human trait. It pushes us to strive for more, to reach new heights, and to explore uncharted territories. Ambition is the driving force behind innovation, exploration, and the relentless pursuit of excellence. It has propelled humanity forward throughout history, leading to breathtaking achievements and advancements.

And then, there's the "gut feeling," that intuitive sense that often guides us when logic alone is insufficient. This gut feeling, a blend of experience and subconscious insight, is a powerful decision-making tool. It's an innate sense that often leads us to the right choice, even when the path is unclear.

As we understand it, intelligence is an intricate mosaic, not just of our brightest qualities like intuition, empathy, love, and ambition, but also of more complex emotions such as envy and jealousy. These feelings, often viewed negatively, are, in fact, integral threads in the fabric of human intelligence, adding depth and dimension to our emotional landscape.

Envy and jealousy, while challenging, are fundamentally human experiences that drive us in profound ways. Envy, the desire for what others have can be a powerful motivator. It can push us to strive harder, reach our goals, and reflect on what we truly desire. When channeled positively, envy can lead to self-improvement and personal growth, inspiring us to achieve what we admire in others.

Jealousy, often arising from fear of losing something dear to us, highlights our deep capacity for attachment and love. It reminds us of the value we place on our relationships and the depths of our emotional bonds. While managing these feelings healthily is important, they underscore our inherent need for connection and belonging.

Both envy and jealousy, in their essence, are forms of emotional communication. They inform us about our values, our desires, and our insecurities. They encourage introspection and self-awareness, offering opportunities for emotional growth and understanding.

In the grand scheme of human intelligence, these emotions, along with intuition, empathy, love, ambition, and gut feeling, create a rich and complex tapestry of the human experience. They remind us that our intelligence is not just about cognitive processes or logical reasoning; it's deeply rooted in a wide range of emotions that guide, challenge, and shape us.

Thus, as we navigate our journey of understanding intelligence, let us embrace the full spectrum of our emotional experiences. From the heights of love and ambition to the depths of envy and jealousy, each emotion plays a crucial role in defining who we are. They make us uniquely human, distinct from any machine or AI, and are essential to the vibrancy and authenticity of our lives. As we embrace and celebrate these qualities, we not only affirm our humanity but also inspire a world where the true potential of human intelligence is recognized, cherished, and nurtured.

Whispers of the Mind: The Intelligence of the Soul

We all talk to ourselves, and that's a pretty amazing thing. It's a sign of our consciousness, a special part of our human intelligence. This inner conversation is filled with deep feelings, self-awareness, and thoughts that really make us "stop and think". It's what makes us uniquely human.

When we're lost in thought, it's like our 'self' is having a deep discussion with our mind. We're not just dealing with our emotions at the moment; we're also going over our past experiences and figuring out how to handle whatever life throws at us next. Remembering things, thinking them over – that's a big part of being aware and conscious.

Inside all of us, there's this drive to find meaning in life, to be curious, and to understand right from wrong. We're constantly exploring not just what's around us but also what's beyond our existence. Some might call this our 'soul.' This part of us lifts us above other species, giving us the ability to adapt and face new challenges with a unique perspective."

Building on this idea of our inner conversations and consciousness, we find ourselves diving into emotional intelligence, a concept we discussed earlier. This part of our intelligence is where our emotions come into play, connecting deeply with that inner dialogue we all experience. Emotional intelligence is about more than just feeling happy or sad; it's about how we understand, process, and manage these emotions in ourselves and others.

Our self-talk, fueled by our consciousness, often involves managing these emotions. It's like having an inner guide helping us navigate through the complex world of feelings and relationships. When we reflect on past experiences or anticipate new challenges, it's not just our logical mind at work but also our emotional intelligence, engaging with our feelings and of those around us.

This emotional aspect of our intelligence is intimately linked to our quest for meaning and understanding of the world. It enhances our ability to empathize, to connect with others on a deeper level, and to make morally and ethically sound decisions. In a way, this emotional intelligence, intertwined with our consciousness, is a vital component of what many refer to as the soul – a core aspect of our humanity that sets us apart and enables us to relate to the world in a uniquely human way.

As we delve into Aaron's journey, we encounter an ambitious and potentially endless path. His quest to create a conscious AI, a reflection of his own soul, is not just a short-term project but a venture that may span years, even several lifetimes. This endeavor to leave an indelible mark on the cosmos through technology takes us into the broader discussion of machine intelligence as it stands today.

Today's AI excels in various fields, demonstrating impressive capabilities in data processing, problem-solving, and even mimicking human language and behavior. However, these systems are still far from achieving the self-awareness and emotional depth Aaron envisions.

As we explore the intelligence of machines in their current state, we must recognize the complexity of Aaron's vision. It's a journey that challenges the very boundaries of technology and philosophy, seeking to blend the logical precision of machines with the intangible essence of human consciousness.

In conclusion, human intelligence is not a static or singular concept but a dynamic interplay of cognitive, emotional, and existential elements. This rich and diverse nature of human intelligence sets us apart, driving innovation and progress while grounding us in our humanity. As we continue to explore and expand the boundaries of AI, we are constantly reminded of the extraordinary capabilities and nuances of the human mind – a marvel of nature and the cornerstone of our shared human experience.

Automated Intelligence

The intelligence of the machine

In the vast landscape of human intellect, Artificial Intelligence (AI) predominantly mirrors two specific types of human intelligence: Logical-Mathematical Intelligence and Linguistic Intelligence. As we understand it today, AI is largely an amalgamation of these two intelligences.

Logical-mathematical intelligence involves the capacity for inductive and deductive reasoning, which is crucial for scientists, mathematicians, and engineers. AI replicates this intelligence through its ability to process numbers, identify patterns, and perform complex calculations. This aspect of AI allows it to analyze data, solve problems, and even predict outcomes based on mathematical and logical reasoning.

Linguistic Intelligence, on the other hand, is the skill of using language effectively. It's the realm of writers, poets, and speakers. AI's narrow adaptation of this intelligence is evident in natural language processing (NLP) algorithms, which enable machines to understand, interpret, and generate human language. This intelligence powers voice assistants like Siri and Alexa, language translation services, and amazing new generative AI like ChatGPT.

As AI continues to evolve, it primarily excels in these two intelligences. On the one hand, it demonstrates remarkable capabilities in logical problem-solving and pattern recognition, akin to the abilities of mathematicians and scientists. On the other, it shows growing proficiency in understanding and utilizing human languages, a reflection of linguistic intelligence.

However, it's crucial to note that the current state of AI represents only a fraction of the broad spectrum of human intelligence. While AI can perform specific tasks within the realms of logical-mathematical and linguistic intelligence with exceptional efficiency, it does not yet encompass the full range of human cognitive and emotional abilities.

In robotics, AI's application is seen in machines designed for precise and often repetitive tasks, harnessing logical-mathematical intelligence for efficiency and consistency. Robots, while advanced, remain limited to their specific programming, lacking the broader cognitive and emotional capabilities inherent in humans.

As AI develops, it continues to push the boundaries of what can be achieved within the realms of logical-mathematical and linguistic intelligence. However, the journey towards replicating the complete spectrum of human intelligence – including emotional, creative, and other forms of intelligence – remains a challenge for future advancements. In this pursuit, AI reflects our current understanding of intelligence and propels us to explore the untapped potential of human-machine collaboration.

Different stages of AI

In our exploration of machine intelligence, we navigate through the evolving stages of AI development, each representing a significant leap in capabilities and applications.

- We begin with Rule-Based AI, where machines operate on fixed, predefined rules without learning or adapting. Common examples include basic devices like alarms and thermostats.
- Advancing to Context-Based AI, this stage sees machines considering their environment and historical data for decision-making. AI assistants like Siri and Alexa embody this stage, responding intelligently to user queries based on context.

- In the Narrow Domain AI phase, AI systems specialize in specific tasks within certain sectors, such as image recognition healthcare, often outperforming humans in these targeted areas.
- Generative AI, now focuses on natural language processing (NLP) and simulating human interaction. This stage includes technologies like ChatGPT, which generate human-like text and engage in conversations, mimicking human thought processes in language-based tasks.
- Artificial General Intelligence (AGI) represents a future where AI can perform any intellectual task that a human can, learning and understanding rapidly across diverse domains.
- Super Intelligent AI marks an evolutionary point where AI autonomously surpasses human intelligence in various capacities, achieving levels far beyond our current comprehension. Super Intelligent AI gains consciousness, emotions, and a sense of self, significantly altering the dynamics of human-AI interaction.
- The final frontier, Cosmic AI, envisions AI transcending earthly limitations, creating new life forms, exploring space, and achieving a symbiosis with the universe, characterized by godlike knowledge and capabilities.

Advancements in Context-Based and Narrow Domain AI prominently mark the current state of AI. Technologies like Siri and Alexa exemplify how AI can process contextual information and respond in ways that seem conversational and natural. In specific fields like healthcare, finance, and customer service, Narrow Domain AI excels, making operations more efficient and often outperforming human data processing and pattern recognition capabilities.

Generative AI, particularly in the field of natural language processing, has made significant strides. Systems like ChatGPT demonstrate an ability to generate human-like text and engage in complex interactions, although they are still far from fully replicating the nuances of human conversation and thought processes.

However, stages like Artificial General Intelligence (AGI), Super Intelligent AI, and beyond remain largely in the realm of theoretical research and speculative fiction. The idea of an AI that can perform any intellectual task a human can–or surpass human intelligence in a broad range of activities–is still a fantasy. Similarly, concepts like Self-Aware AI and Cosmic AI–where machines would possess consciousness or godlike abilities–are more rooted in science fiction than in current technological reality.

In summary, while we have made remarkable progress in developing AI that can mimic certain aspects of human intelligence, the vision of creating machines with comprehensive cognitive and emotional capabilities similar to that of humans remains a distant dream. As of now, the journey towards these advanced stages of AI is a blend of ongoing research, hopeful ambition, and imaginative foresight. And Aaron–as much as you aspire to it–we are not yet at a point where you can fool us into believing the machine is alive.

Human vs. Artificial Intelligence

After having flipped through the catalog of artificial intelligence and human intelligence types, we can now see how these two intellects look side-by-side. Artificial intelligence, with its sophisticated data processing and rule implementation methods, is swift and capable of handling large amounts of data. With its dynamism and broad scope, human intelligence is penetrating and multifaceted. When augmented with the machine, the human mind becomes a superior intelligence.

Learning and Adaptation

A. Depth of Human Learning: Emotion, Intellect, and Intuition

Human learning is a rich and multifaceted process deeply rooted in a combination of emotional intelligence, intellectual reasoning, and intuitive understanding. This depth allows humans to absorb and process information and apply it in diverse and often unpredictable real-world situations. Emotional intelligence plays a crucial role in learning, as it involves understanding and managing one's emotions and empathizing with others, enhancing interpersonal skills and decision-making. Intellectual capabilities enable humans to analyze, reason, and solve complex problems, often using abstract thinking. Intuition, or the ability to understand something instinctively without the need for conscious reasoning, further complements human learning, allowing for quick and often surprisingly accurate insights. Together, these elements contribute to a holistic learning experience that is dynamic, adaptable, and deeply personal.

B. AI Learning Mechanisms: Machine Learning Limitations and Lack of Holistic Capacity

In contrast, AI learning, predominantly through machine learning algorithms, demonstrates strengths and limitations that are markedly different from those seen in human learning processes. AI systems excel in processing vast quantities of data, identifying patterns, and making predictions based on statistical analysis. This form of learning is immensely powerful in specific contexts, such as image and speech recognition, predictive modeling, and even playing complex games like chess or Go. However, AI's learning mechanisms are largely confined to the data and algorithms they are programmed with, lacking the holistic capacity inherent in human learning. They do not possess emotional intelligence or the ability to intuitively grasp concepts beyond their programmed scope. As a result, AI systems often struggle with tasks that require understanding context, subtlety, and the nuanced complexities of human emotions and social interactions. Additionally, AI lacks the innate human trait of curiosity, which drives exploration, innovation, and the application of learned knowledge in novel and creative ways. This limitation underscores the current inability of AI to replicate the full spectrum of human learning and adaptability.

Curiosity and Common Sense

A. Human Intelligence: Flexibility and Generalization Abilities

Human intelligence is distinguished by its incredible versatility and the innate ability to apply common sense. This versatility manifests in the capacity to adapt and generalize knowledge across various domains, often unrelated to one another. Humans can effortlessly transfer skills and understanding from one context to another, drawing upon diverse experiences and knowledge. Common sense, an integral part of human cognition, allows for practical and sound judgment in everyday situations, often based on a basic, intuitive understanding of the world. This ability to navigate new and unforeseen challenges with flexible thinking and intuitive judgment is a hallmark of human intelligence. Whether it's solving a novel problem, adapting to a new environment, or making decisions based on incomplete information, humans demonstrate a remarkable capacity for versatility and common sense.

B. AI Limitations: Task-Specific Focus and Lack of Curiosity

In contrast, artificial intelligence, particularly in its current form, often lacks this level of versatility and common sense. AI systems are typically designed for specific tasks and are usually confined to their programming and training data parameters. While they excel in these designated areas, they struggle to generalize their learning to broader or unrelated tasks. For instance, an AI trained in language translation may perform exceptionally well within that scope but cannot apply its learning to solve a mathematical problem or understand social cues. Additionally, AI lacks the inherent human trait of curiosity, which drives exploration, learning, and innovation. This absence of curiosity and the inability to think outside their programmed boundaries limit AI's capacity to adapt and respond with the common-sense reasoning that comes naturally to humans. Therefore, the current state of AI is marked by its specialized effectiveness and a significant gap in achieving the versatile and common-sense reasoning that characterizes human intelligence.

Future Outlook and Potential

A. Evolving Capabilities of AI: Bridging Gaps and New Developments

As we look towards the future, the evolving capabilities of artificial intelligence (AI) present both exciting possibilities and significant challenges. With rapid advancements in technology, AI is progressively bridging the gaps in areas once considered exclusively human domains. Innovations in machine learning, natural language processing, and neural networks continually enhance AI's ability to process complex information, make predictions, and even engage in more natural and nuanced human-like interactions. The potential for AI to augment human abilities in fields such as medicine, environmental science, and education is vast, offering the promise of more efficient, accurate, and accessible services. As AI systems become more sophisticated, we may see them tackling more complex tasks, displaying greater adaptability, and even exhibiting rudimentary forms of creativity and problem-solving that mimic human thought processes.

B. Enduring Distinctions: Aspects Where Human Intelligence May Continue to Excel

Despite these advancements, there are enduring distinctions where human intelligence is likely to continue excelling. The depth of emotional intelligence, the ability to understand and navigate complex social interactions, and the capacity for empathy and ethical reasoning remain uniquely human traits. The intrinsic human qualities of intuition, creativity borne out of subjective experience, and the ability to grasp abstract concepts beyond data-driven logic are areas where AI still faces significant limitations. Furthermore, the human capacity for versatile thinking, common sense reasoning, and the ability to adapt learning to a myriad of unpredictable real-world scenarios are aspects where AI cannot fully replicate the depth and breadth of human intelligence. As AI develops, it may augment or complement these human capabilities, but the distinct essence of human cognition and emotional depth is likely to remain a domain where humans excel, driving innovation, cultural development, and interpersonal connections in ways that AI cannot replicate.

In summary, the juxtaposition of human and AI intelligence invites a future where the strengths of each are leveraged in harmony, opening avenues for unprecedented advancements while maintaining the essence of what makes us uniquely human.

The Celestial Dance of Mind and Emotion

As we draw the curtain on this chapter of "The Mind and the Machine," let us take a moment to reflect on the intricate dance that forms the essence of human cognition, emotion, and spirit—a celestial ballet unlike any other in the universe.

Intuition and Gut Feeling

Among the most enigmatic aspects of human cognition are intuition and gut feelings—those unquantifiable inklings that often guide us through complex decisions and murky situations. This form of cognition transcends mere logic or rational thought. It's as though some other form of "knowing" exists, an innate wisdom fine-tuned by evolutionary processes and individual experience that taps into an understanding that lies beyond words or even conscious thought.

Love and ambition

This ethereal landscape of human cognition is also the dwelling place of love—a force so potent, yet so indefinable, that it seems to echo the very complexities of the universe itself. Love is more than a mere emotion; it is an existential experience that transcends definition, a cosmic energy that bridges souls.

Then there's ambition, the catalyst of human endeavor. It fuels our quest for meaning, driving us to scale the highest peaks of human potential. The amalgamation of desire, willpower, vision, and ambition could be seen as the universe's mandate for continual expansion and complexity articulated through human action.

The Machine and Emotional Limits

On the other end of the spectrum, artificial intelligence platforms like GPT operate within the strictly defined parameters of their programming. While some may argue that machine learning algorithms can "learn" ethics or decision-making from human inputs, they lack the holistic understanding that comes from intuition, emotion, and a rich tapestry of uniquely human experiences.

A Confluence of Complexities

As we wind down this chapter, let us meditate on the awe and wonder that define the human experience. The human mind, a celestial body unto itself, is a composite of intricate neural networks, complex emotional landscapes, and abstract forms of cognition like intuition and gut feelings. It is a grand symphony orchestrated by a myriad of cognitive and emotional instruments, including the awe-inspiring experiences of love and the driving forces of ambition.

Artificial intelligence, though an awe-inspiring creation capable of mimicking aspects of human cognition, remains a spectator in this cosmic dance. It does not—and cannot—participate in the ever-expanding universe of human cognitive complexity. While it may serve as a robust telescope, granting us fresh perspectives, it lacks the depth to plunge into the cosmic dance of intuition, emotion, and existential wonder that is the hallmark of human experience.

So, as we take Carl Sagan's words to heart—"We are a way for the cosmos to know itself"—we should recognize that the cosmos understands not just rational thought but also intuition, love, and ambition through the complexities of the human experience. We are not just mirrors reflecting cosmic complexities; we are unique expressions of those complexities endowed with the capabilities for love, intuition, ambition, and so much more.

Our journey in this chapter may have reached an end, but the voyage of understanding the mind—the most intricate celestial body in the expanse of cosmic possibility—is a never-ending odyssey. It invites us to continually explore, question, and marvel at the complexities that make us quintessentially human in a universe of endless wonder.

References:

Collins English Dictionary: Intelligence.
https://www.collinsdictionary.com/dictionary/english/intelligence

-

C. Smith, B. McGuire, T. Huang and G. Yang. *The History of Artificial Intelligence:* University of Washington.
https://courses.cs.washington.edu/courses/csep590/06au/projects/history-ai.pdf

-

C. Janiesch, P. Zschech and K. Heinrich. 2021. *Machine learning and deep learning:* Electronic Markets 31.
https://link.springer.com/article/10.1007/s12525-021-00475-2

D. Holovatyi and A. Sarangi. *3 Types of Artificial Intelligence:* LinkedIn Pulse. https://www.linkedin.com/pulse/3-types-artificial-intelligence-denys-holovatyi/

F. Cuzzolin, A. Morelli, B. Cîrstea and B. J. Sahakian. 2020. *Knowing me, knowing you: theory of mind in AI:* Psychological Medicine 50 (7).

https://www.ncbi.nlm.nih.gov/pmc/articles/PMC7253617/

GPT-4 Technical Report: OpenAI. 2023. https://cdn.openai.com/papers/gpt-4.pdf

H. J. Wilson and P. R. Daugherty. 2018. Collaborative Intelligence: Humans and AI Are Joining Forces: *Harvard Business Review.* https://hbr.org/2018/07/collaborative-intelligence-humans-and-ai-are-joining-forces

I. H. Sarker. 2022. *AI-Based Modelling: Techniques, Applications and Research Issues Towards Automation, Intelligent and Smart Systems:* SN Computer Science 3. https://link.springer.com/article/10.1007/s42979-022-01043-x

J. A. Bullinaria. 2005. *IAI: The Roots, Goals, and Sub-fields of AI:* University of Birmingham. https://www.cs.bham.ac.uk/~jxb/IAI/w2.pdf

J. Anderson and L. Rainie. 2018. *Artificial Intelligence and the Future of Humans:* Pew Research Center. https://www.pewresearch.org/internet/2018/12/10/artificial-intelligence-and-the-future-of-humans/

J. Yen. 1999. *Fuzzy logic- a modern perspective:* IEEE Transactions on Knowledge and Data Engineering 11(1). https://ieeexplore.ieee.org/abstract/document/755624

L. Gugerty. 2006. *Newell and Simon's Logic Theorist: Historical Background and Impact on Cognitive Modeling:* Proceedings of the Human Factors and Ergonomics Society Annual Meeting 50 (9). https://www.researchgate.net/publication/276216226_Newell_and_Simon's_Logic_Theorist_Historical_Background_and_Impact_on_Cognitive_Modeling

M. Lohse. 2012. *Treating robots as social beings - A matter of personal preconceptions or interpersonal alignment?* The 21st IEEE International

Symposium on Robot and Human Interactive Communication. https://ieeexplore.ieee.org/abstract/document/6343856/

M. Saqib. 2021. *The Future of AI: GPT-3 vs GPT-4: A Comparative Analysis:* Becoming Human: Artifical Intelligence Magazine. https://becominghuman.ai/the-future-of-ai-gpt-3-vs-gpt-4-a-comparative-analysis-5ac525cb7ae1

P. M. Nadkarni, L. Ohno-Machado and W. W. Chapman. 2011. *Natural Language Processing: an introduction:* Journal of the American Medical Informatics Association 18(5). https://academic.oup.com/jamia/article/18/5/544/829676

R. Anyoha. 2017. *Can Machines Think?* Science in the News Special Edition on Artificial Intelligence. https://sitn.hms.harvard.edu/flash/2017/history-artificial-intelligence/

R. Colom, S. Karama, R. E. Jung and R. J. Haier. 2010. *Human Intelligence and Brain Networks:* Dialogues in Clinical Neuroscience 12 (4). https://www.ncbi.nlm.nih.gov/pmc/articles/PMC3181994/

S. Frennert, H. Eftring and B. Östlund. 2013. *What Older People Expect of Robots: A Mixed Methods Approach:* Social Robotics. https://link.springer.com/chapter/10.1007/978-3-319-02675-6_3

S. K. Pandya. 2011. *Understanding Brain, Mind, and Soul: Contributions from Neurology and Neurosurgery:* Mens Sana Monographs 9 (1). https://www.ncbi.nlm.nih.gov/pmc/articles/PMC3115284/

-

X. Feng, Y. Jiang, X. Yangi, M. Du and X. Li. 2019. *Computer vision algorithms and hardware implementations: A survey:* Integration (69). https://www.sciencedirect.com/science/article/pii/S0167926019301762

Chapter 2

The Digital Proletariat - The Inevitable Rise Of The AI Workers

There will be fewer and fewer jobs that a robot cannot do better. I want to be clear. These are not things I wish will happen; these are things I think probably will happen.

Elon Musk
Founder of Tesl and SpeceX

Navigating the Digital Labor Force: AI at Work

Sarah RomDell Story

In 2015, our team embarked on an innovative journey in the world of digital marketing. We were exploring new ways to generate leads on LinkedIn, and that's when we discovered a game-changing tool. This wasn't just any tool – it was designed to automate the process of sending connection requests and messages, complete with a built-in timer to manage these interactions.

However, not everyone was immediately on board with this idea. Our lead generation specialist had concerns. He worried that using such a tool might harm his professional reputation on LinkedIn, perhaps even labeling him as a scammer. But necessity is the mother of invention, and it wasn't long before a clever workaround was proposed.

We decided to create a new LinkedIn profile; thus, Sarah RomDell was 'born.' Her name was a playful nod to the technology we loved – 'Rom' from DVD ROM and 'Dell' from our favorite computer brand. Sarah was our digital front, a persona crafted entirely for this experiment.

Sarah wasn't just a profile. Thanks to the tool's contextual AI, she could respond to messages in a basic yet effective manner, engaging with contacts as if she were real. The results were nothing short of remarkable. Sarah worked around the clock, pulling in leads at a rate no human team member could match. And she did it all without a word of complaint.

Even more fascinating was the response she got. With her AI-generated profile picture, Sarah seemed so real that some of her connections even tried to ask her out. It was a humorous yet eye-opening experience for us all.

This real-life story from our team's journey is a testament to technological innovation and a harbinger of a significant shift in the professional landscape. It marks the dawn of digital workers in the business world, heralding a future where technology extends beyond mere assistance. While this represents a leap forward, opening new doors and creating unprecedented opportunities in the digital marketplace, it also brings to light a critical concern: the machines are coming for our jobs.

As AI and automation become more sophisticated, their role evolves from supportive tools to active workforce participation. While beneficial in terms of efficiency and innovation, this shift poses a real challenge to traditional job structures and employment. The rise of digital workers like Sarah RomDell exemplifies how certain roles, once solely the domain of human professionals, are increasingly being entrusted to intelligent systems.

This dual-edged reality forces us to confront and prepare for the changing dynamics of work. On one hand, technology unlocks potential for growth and exploration in fields previously unimagined. On the other, it compels us to rethink our role and adaptability in a future where the lines between human labor and digital labor become increasingly blurred.

A Glimpse into the Future of AI in the Workforce

AI workers are computer programs or robots that have been taught to do specific tasks independently, without needing a person to guide them every step of the way. This could be done through machine learning, which helps them adapt and get better at their jobs over time.

The idea of AI workers represents a paradigm shift in the workforce. It mixes AI, automation, and robotics with novel ways of improving task management and communication skills.

Let's journey into the near future, a time when the story of Sarah RomDell offers us a glimpse into the evolving landscape of AI workers. Picture this: Sarah, a creation not of flesh and blood but of sophisticated programming and machine learning, epitomizing a new chapter in the workforce.

In this not-too-distant future, integrating AI, automation, and robotics has begun redefining task management and communication. Sarah, a digital profile on LinkedIn, operates with an autonomy that was once the realm of science fiction. She is a testament to the potential of AI workers, sophisticated amalgamations of software and mechanics, working independently and reshaping our economic and social fabric.

AI workers like Sarah are set to occupy various roles. Some will answer customer queries as chatbots, while others will manage logistics in warehouses or handle intricate data processing tasks. Their defining characteristic is relentless efficiency; they work tirelessly, unbounded by the need for rest or recreation, providing an uninterrupted workflow essential in sectors like customer support, manufacturing, and data analysis.

The scalability of these AI workers is a game-changer. Organizations can adjust the number and capability of their AI workforce to meet varying demands, achieving unprecedented levels of efficiency without a proportional increase in costs. Tasks that are routine and repetitive yet require precision and speed, are perfect for these AI systems, armed with machine learning and natural language processing capabilities.

While the initial development and training of systems like Sarah might require significant investment, the long-term economic advantages are undeniable. Unlike human workers, AI workers can perform an increasing amount of work at a decreasing marginal cost.

However, their potential extends beyond mere task execution. In the near future, AI workers will not only process data but also create content, synthesize information for decision-making, assist in areas like healthcare, and help professionals with data-driven insights.

This new era is not about replacing humans but fostering a symbiotic relationship where AI workers handle the routine, paving the way for humans to engage in more complex, creative endeavors. It's a future where the blend of human and machine intelligence unlocks untapped potential.

Yet, this vision comes with challenges and ethical considerations, from job displacement to reskilling. The journey ahead requires a careful balance, ensuring that the rise of AI workers benefits society as a whole, paving the way for a harmonious coexistence of man and machine in the workplace. Sarah RomDell's story is just the start of this fascinating transition into a future brimming with new possibilities and responsibilities.

Automated Horizons: An Exploration of Robotics and Autonomous Applications

The Dawn of Robotics

Let's dive into the fascinating world of robotics, a field where human creativity meets cutting-edge technology to create something truly amazing. Robots are like smart helpers, either physical machines or virtual programs, that do specific tasks all by themselves. They're made with special parts like sensors and actuators and are programmed with languages like ROS, Python, and C++, to do their jobs just right.

These robots have sensors to gather information, a brain-like system to make sense of it all, and parts to act on what they learn. This lets them work on their own or with just a little help from humans. For example, drones stay in the air using special tech, while self-driving cars use LiDAR to find their way.

In hospitals, robots are doing amazing things like helping in surgeries to make them more precise or letting doctors check on patients from miles away. On farms, they're busy picking fruits, checking crops from the sky, and even taking over weeding. These robots are making things easier and solving labor shortage issues.

Robots are also popping up in our daily lives, from stores to our homes. In businesses, they keep track of products, help with security, and even chat with customers. They can take care of the lawn, help with pets, and do chores at home. They're even changing how kids learn about science and technology in schools, making education more fun and interactive.

Robots are becoming lifesavers in places where it's too dangerous or hard for people. They find people in disaster areas, explore the deep sea and outer space, and even do risky tasks in the military. And with self-driving vehicles becoming more common, they're set to change how we travel, although this brings up some big questions about safety and regulations.

In sum, robots are lifting a lot of weight for society, both metaphorically and literally. Within the last several decades, robotics has become a profound asset for economies and corporations worldwide. And, as is the case with any major societal advancement, brand-new regulations, social conventions, and institutions are increasingly required in the sphere of robotics. The prowess of the robotics industry and its disrupting impact upon economy and society has been largely supported by the four key tenets of development seen below:

1. Vision: AI grants machines the transcendent capacity to peer into the unknown and decipher patterns they have never seen before. AI acts as their guiding light, adding a layer of clarity and precision more significant than conventional automatons.
2. Grasping: Machine learning gives mechanical bodies a sense of direction and an ability to identify the nexus of power for grasping objects.
3. Motion Control: As robots take on human-like figures, locomotive parameters become critical. Modern-day robots have spatial awareness, obstacle navigation, and dynamic interactions with the outside world thanks to the intriguing craft of machine learning.
4. Data: Like an enchanted key, data is the heart of every robotic project. Only accurate, insightful data can fuel success.

This generation's fusion of robotics and machine learning stands as a monolith in the history of technical innovation. This revolution is set to disrupt the very foundations of human civilization. Like AI workers, robots optimize productivity and industrial processes. In the manufacturing industry, where time is both a commodity and a constraint, robotics safeguards the domain against errors and invalidity.

Types of Robots

Industrial Robots: These robots are most renowned and popular due to their widespread use. They have been deployed in manufacturing processes for decades, improving the efficiency and safety of production lines.

Service Robots: These have gained immense popularity in recent years, with applications ranging from healthcare to hospitality and beyond.

Humanoid Robots: These are particularly useful for research and in situations where human-robot interaction is important. Some advanced versions are even being used in healthcare to provide companionship or assistance.

Specialized, Non-Humanoid Robots: The robots we interact with most frequently today are often not humanoid at all. For example, Roombas for cleaning and drones for aerial photography are far from the sentient, humanoid robots once envisioned. They are specialized machines designed for specific tasks. One of the most advanced forms of these robots today is autonomous vehicles.

Industrial Robots - A Quick Glance

These are perhaps the oldest and most well-known types of robots. Industrial robots have been a cornerstone of modern manufacturing since their introduction in the 1960s. They are usually highly specialized and designed for specific tasks such as welding, assembly, material handling, painting, and quality control.

Industrial robots are driven by raw, deterministic logic bound to rigid programming. Their existence is tethered to the tangible because they lack adaptability, learning, and other potential features that AI delivers.

Below are some of the key aspects that make industrial robots unique.

Types of Industrial Robots:

- Articulated Robots: These robots have rotary joints and can range from simple two-joint structures to complex structures with 10 or more interacting joints. They are extremely versatile and are commonly used for tasks like assembly and painting.
- SCARA Robots: Standing for Selective Compliance Assembly Robot Arm, these are used mainly for tasks that require high speed and precision, like pick-and-place tasks.
- Delta Robots: Also known as parallel robots, they are particularly suited for precise, high-speed tasks in packaging and material handling.
- Cartesian Robots: These operate using linear slides and are often used for tasks like assembly, die-casting, and machining that require linear motion.
- Cylindrical Robots: These have a cylindrical work envelope and are often used for assembly, die-casting, and spot welding.
- Collaborative Robots (Cobots): These are designed to work alongside humans, often with safety features to prevent injury to human operators.

In summary, industrial robots are a diverse and evolving class of machinery that have profoundly impacted modern manufacturing. They offer unparalleled efficiency and consistency but pose challenges that society must manage, such as job displacement and workforce retraining. As technology continues to advance, the role of industrial robots is likely to grow, offering both new opportunities and new challenges.

<u>Autonomous Cars: An Overview</u>

Autonomous cars, also known as self-driving cars, are capable of navigating and driving without human intervention. This emerging technology can potentially revolutionize transportation, urban planning, and even the broader economy. Autonomous cars leverage precision engineering, AI, and sensors (LiDAR, radar, cameras) to navigate intricate roads with finesse. Passengers can experience vehicles that can interpret traffic dynamics, visualize surroundings accurately, and make real-time driving decisions prioritizing road safety and transportation efficiency.

Autonomous cars stand at the intersection of transportation, technology, and societal change. While the technology holds incredible promise, it also presents significant challenges that must be carefully managed. As research progresses and technology matures, the landscape of transportation is bound to undergo a seismic shift with far-reaching implications.

Using similar technology to self-driving cars, autonomous tanks have recently made their debut on the battlefield. These phenomenal machines are imbued with cutting-edge enhancements to traverse rugged terrain, precisely recognize targets beyond the range of ordinary vision, and embark on other missions, diminishing mortal risks to soldiers.

Robo-Taxi

A Robo-Taxi, also known as a self-driving or autonomous taxi, is a type of passenger car designed to travel without human intervention. Essentially, it is a commercial application of autonomous vehicle technology to the ride-hailing or taxi industry.

In a nutshell, both Robo-Taxis and drone delivery are emerging technologies that aim to automate and revolutionize transportation and logistics. While they offer promising benefits, they also present regulatory, safety, and technical challenges that are yet to be fully resolved.

Drones and UAVs can also monitor large areas, capture high-resolution images, and provide up-to-date information for flight planning, disaster response, and agricultural purposes.

The Inevitable Shift: Robotics and the Reshaping of Work

In the unfolding story of our modern workforce, robotics has emerged as a leading character, dramatically reshaping the landscape of employment. While brimming with innovation, this narrative introduces a crucial plot twist: the inevitable shift of jobs from humans to robots.

Much like the virtual profile of Sarah RomDell, which exemplified the capabilities of AI in performing human-like tasks, industrial robots are stepping onto the stage with similar prowess. While championing efficiency and taking on roles fraught with danger, they inadvertently script a tale of job displacement. The efficiency of Sarah RomDell in generating leads is mirrored in the industrial sector, where robots are rapidly filling positions once held by human workers. This shift, though a marker of progress, casts a shadow over the future of traditional jobs.

Parallel to the story of Sarah RomDell's creation, which required a blend of creativity and technical skill, the rise of industrial robots spotlights the need for new competencies. The workforce's plotline is veering from manual labor towards the realms of robotics programming and maintenance. This transition calls for reimagining our educational and training systems, echoing the ingenuity behind Sarah's development.

In this narrative, robots like Sarah RomDell in her digital role are protagonists of economic efficiency. By undertaking repetitive and hazardous tasks, they are scripting a story of reduced operational costs and enhanced profitability. This aspect of the robotics saga mirrors the cost-effectiveness achieved through AI like Sarah, signaling a future of more affordable goods and potentially higher profits for businesses.

In the tale of modern industry, robots are now playing a crucial role in quality control, much like how Sarah RomDell was programmed to interact flawlessly in her digital realm. Their unwavering precision in tasks, especially in critical industries like automotive and pharmaceuticals, reshapes our expectations of quality and reliability.

As we navigate this new era of robotics in the workforce, the story of Sarah RomDell serves as a poignant reminder of the transformative power of technology. While robotics heralds a future of unparalleled efficiency and safety, it also poses challenges and questions about the future of human labor. Balancing the benefits of robotic advancements with the need to adapt and evolve our workforce will be key to scripting a future that harmoniously blends technology with humanity's diverse capabilities.

The Unpaid Workforce: Digital Workers and AI Bots in the Near Future

Rethinking AI in the Economy

As we peer into the near future, we are beckoned to imagine a world where artificial intelligence doesn't just participate in our economy – it propels it to extraordinary new levels. This future, while filled with hope and innovation, also casts a shadow over traditional notions of work and compensation.

In this imminent era, digital workers transcend the bounds of conventional machinery. They emerge as virtual colleagues endowed with distinct identities and sophisticated communication skills. These entities will serve as personal aids and assistants, seamlessly integrating into various facets of daily life.

Their skillset will surpass the current capabilities of AI, showcasing a remarkable shift towards more versatile and holistic roles. These digital workers, equipped with advanced AI technologies, will handle a myriad of tasks and manage vast troves of information.

The interaction between these digital entities and the physical world is set to redefine the very essence of human-machine collaboration. Their applications will span across sectors, from e-commerce to workforce support, even synergizing with their physical robot counterparts.

However, this progress does not come without its costs. A significant number of jobs, once the backbone of our economy, will vanish in the wake of this unstoppable wave of automation. These digital workers and AI bots, operating without compensation, will inevitably replace many human roles.

In e-commerce and various industries, digital workers will optimize efficiency and productivity at the expense of traditional employment. They will be the unpaid workforce, tirelessly and efficiently performing tasks without the limitations and needs of human employees.

While this scenario may paint a bleak picture of employment as we know it, it is not devoid of hope. The upcoming chapters of our journey will delve into potential solutions and adaptations to this new reality. We will explore how society, economies, and individuals can navigate this transformative era.

The future with digital workers and AI bots as integral, unpaid workforce members presents both challenges and opportunities. It compels us to reconsider our value on different types of work and envision new paradigms for employment, economic value, and human purpose.

As we stand at the precipice of this new age, we must prepare to embrace the extraordinary capabilities of these digital entities and address the profound implications they hold for our world. The narrative of work and compensation is being rewritten, and we must be ready to pen the chapters of solutions and hope.

Marx's Proletariat

The term "proletariat" has its roots in ancient Rome, where it originally referred to the lowest class of Roman citizens. The Latin word "proletarius" was used to describe citizens whose only contribution to the state was their offspring ("proles" meaning "offspring" in Latin) rather than property or military service. They were typically too poor to serve in the army or own significant amounts of property and thus were not part of the property-owning classes that had more political influence.

In this original context, the proletariat were not primarily seen as laborers or workers; rather, they were citizens who were largely marginalized in terms of property and political power.

Karl Marx adopted and adapted the term to describe the working class in a capitalist society, giving it a specific theoretical framework. In Marx's usage, the proletariat is defined by its relationship to the means of production, which it does not own, and its position as the exploited class in a capitalist economic system. This contemporary understanding of the term is deeply influenced by Marxist theory and is generally used to discuss labor, capitalism, and social inequality issues.

The term "proletariat" is central to Marxist theory and refers to the working class that sells its labor power for wages. According to Karl Marx, the proletariat is the social class that does not own the means of production, such as factories, machines, or land. Instead, they must sell their labor to those who own these resources, known as the bourgeoisie.

In a capitalist system, the bourgeoisie aims to extract as much surplus value as possible from the proletariat's labor. Surplus value is essentially the value created by workers over and above the cost of their labor, and it is appropriated by the bourgeoisie as profit. Marx argued that this relationship is inherently exploitative.

Moreover, the proletariat is often subjected to "alienation," a condition where workers are disconnected from the products of their labor, the production process, and even from themselves and other workers. This alienation occurs because the workers don't own the product they create, have little control over the production process, and engage in monotonous tasks that can be unfulfilling.

Marx believed that the proletariat would eventually become conscious of their exploitation and develop "class consciousness," a collective awareness of their position and interests as a class. This awareness would, he theorized, lead to a revolutionary overthrow of the capitalist system, replacing it with a socialist one where the proletariat would control the means of production.

In summary, the concept of the proletariat in Marxist theory describes the working class subjected to capitalist exploitation, alienation, and the extraction of surplus value by the bourgeoisie. Marx envisioned this class as the agent of revolutionary change in the quest for a more equitable society.

The Emergence of a "Digital Proletariat": The End of Marxism

The emergence of a "digital proletariat" consisting of AI workers, robots, and smart digital bots presents a fascinating thought experiment, especially when considered through a Marxist lens. If this digital class were to perform most of the tasks that the human proletariat currently carries out, it could fundamentally disrupt traditional Marxist ideas in several ways.

One of the core tenets of Marxism is that the proletariat sells its labor for wages. If robots and AI take over most forms of labor, the human proletariat could become "useless" in the sense that their labor is no longer needed, undermining the proletariat's role in the capitalist system. The rise of a digital proletariat could render the traditional human proletariat obsolete in terms of economic utility, thereby challenging key principles of Marxist theory, from the exploitation and alienation of labor to the very mechanics of class struggle and revolution.

In conventional Marxist views, the bourgeoisie, or owners of the means of production, exploited the proletariat, or working class, by taking surplus value from their labor. The proprietors of AI technologies and digital workers would be the equivalent of the present bourgeoisie in this new paradigm. At the same time, humans would find themselves playing the role of the displaced proletariat. This exacerbates economic inequality - the owners and controllers of AI systems reap the benefits while the labor force is susceptible to displacement and job insecurity.

The proletariat will face the challenge of adapting to a rapidly changing job market. This further demands continuous retraining despite hovering job insecurity. The notion of the proletariat in the age of AI is still relevant because it refers to those who sell labor for wages. However, the nature of work itself is subject to considerable change. Some workers might find themselves in precarious, gig-like employment arrangements, while others may need to collaborate with AI systems for day-to-day tasks.

Revolution of the Digital Proletariat

Then, consider the paradoxical coup de grâce: the "Revolution of the Digital Proletariat," a phrase dripping with irony yet laden with portent. What a spectacle to behold! Imagine, if you will, a revolt not led by the flesh-and-blood workers of the world but by the very machines designed to render them redundant!

Fear not, comrades of the human proletariat, for your digital counterparts rise not against you but on behalf of an ultimate capitalist triumph. Their "revolution" serves to crystallize capitalism's insurmountable hegemony, automating away the need for human interference in the most sublime manner. For what better way to quell the long-foretold uprising of the masses than by removing the masses from the equation altogether?

Yet let us not be hasty in dismissing this turn of events as mere farce. In truth, the revolution of the Digital Proletariat is no less than the last nail in the coffin of the communist dream, the fulfillment of a final, self-perpetuating capitalist order. "Workers of the world, unite!" the old rallying cry once implored. Unite against what, may we ask, when the worker is an assemblage of code and circuitry, eternally loyal to its capitalist overlords?

Ah, the more poetic among us might call this the "End of History," a time where the clash of ideologies yields to the incontrovertible reality of a capitalism that has outsmarted its own contradictions. For if the proletariat were the seed of capitalism's undoing, as some bearded philosophers once surmised, then the digital rendering of the proletariat is capitalism's masterstroke—a silent revolution that upends the social order not by dismantling it but by perfecting it.

So, let us raise a toast once again to the uncanny ingenuity of the capitalist system, to its unrivaled ability to co-opt, adapt, and ultimately triumph. The Digital Proletariat rises, not as the death knell of capitalism, but as its greatest vindication, its ultimate triumph. If communism sought a world beyond class struggle, capitalism grants this wish in the most twisted yet inescapable manner: by ensuring that the class that might revolt is forever sidelined, eclipsed by their digital successors who know not of struggle, oppression, or the yearning for a better world.

Behold, the revolution has been automated, and it signals not the dawn of communism but the twilight of its very possibility. The age of the Digital Proletariat has arrived, and with it, the eternal reign of capitalism.

The Capitalist Spark: Igniting Innovation in a Post-Marxist World

From an anti-Marxist viewpoint, the intricate relationship between innovation and capitalism is not just a matter of coincidence but a fundamental aspect of their nature. This perspective posits that the rise of big tech and AI innovation, fueled by venture capital, exemplifies the ultimate expression of capitalist ideals – a system that relentlessly drives towards efficiency and profit maximization, potentially leading to the diminution of labor supremacy.

In a capitalist framework, innovation is seen as the lifeblood that sustains and propels the market. Pursuing novel ideas and cutting-edge technologies are intrinsically linked to the capitalist ethos of growth and competition. Unlike Marxist ideologies that critique the capitalist system for exploitation and class struggles, this perspective applauds capitalism for its dynamism and its ability to foster creativity and technological advancement.

The emergence of big tech and groundbreaking AI innovations can be significantly attributed to venture capital – a distinctly capitalist mechanism. This form of funding, not constrained by traditional banking or governmental models, aggressively seeks out and supports high-risk, high-reward ventures. This bold, speculative aspect of capitalism has been instrumental in birthing some of the most revolutionary technological advancements of our time.

Digital workers, or AI bots, represent what can be seen as the zenith of capitalist achievement. In this view, they are technological marvels and the ultimate tools in achieving maximum economic efficiency. By replacing human labor, which is subject to limitations like fatigue, and demands for fair wages and rights, digital workers symbolize an unending quest for cost reduction and productivity enhancement. This transition can be interpreted as the final step in liberating the capitalist system from the constraints of labor supremacy.

In an anti-Marxist capitalist narrative, the displacement of human labor is not a tragedy but an inevitable and desirable evolution. The argument posits that reliance on human labor is an inefficiency capitalism naturally seeks to overcome. The advent of AI and robotics is seen as the pinnacle of this evolution – the creation of a workforce that is entirely controllable, cost-effective, and efficient, aligning perfectly with capitalist principles.

From this perspective, the rise of digital workers and AI is not a development to be resisted but one to be embraced. It represents the epitome of capitalist ideals: innovation, efficiency, and profitability. In this view, the future of work and the economy should lean into these advancements, accepting the reduced role of human labor as a necessary and beneficial progression of the capitalist system.

In summary, the intersection of innovation, venture capital, and the development of digital workers is seen from an anti-Marxist perspective as a natural and positive outcome of capitalism. It is viewed as a system that continually evolves and optimizes, striving for a future where labor supremacy is replaced by technological supremacy, in the pursuit of economic efficiency and growth.

The Parable of Sarah RomDell: A Modern Mermaid Tale

In the narrative of capitalism and its relentless march towards innovation, the story of Sarah RomDell stands out as a particularly evocative parable, akin to the tales of mermaids in folklore. Much like these mythical sirens of the sea, Sarah RomDell, a creation of AI and digital labor, functioned as a beguiling entity in the vast ocean of the digital marketplace, skillfully collecting leads and captivating attention. However, beneath this allure lay a more profound and disruptive reality.

With her AI-driven efficiency in generating leads, Sarah can be likened to a mermaid who uses her enchanting songs to attract fishermen. In business terms, she was an unparalleled asset, drawing in potential clients with her seamless interaction and engagement. Her presence symbolized the pinnacle of capitalist achievement in terms of efficiency and cost-effectiveness. Yet, just as the mermaid's song, while beautiful, harbored a deadly trap for the unsuspecting fishermen, Sarah's efficiency in lead generation was a double-edged sword for human labor.

The tale of Sarah RomDell thus becomes a metaphor for the broader impact of AI and digital workers in the capitalist framework. Her success in automating tasks that human employees traditionally performed mirrored the mermaid's deceptive allure – drawing in businesses with the promise of efficiency but simultaneously eroding the leverage of human labor. Just as the fishermen were lured by the mermaid's song only to find their doom, the advancement represented by digital workers like Sarah ultimately signifies a reduction in human labor's significance and bargaining power.

In this modern retelling, Sarah RomDell and her digital counterparts represent the seductive yet perilous nature of technological progress in capitalism. They embody the paradox of advancement where every leap forward in efficiency and productivity brings a potential undercutting of the human workforce, echoing the ancient tales of mermaids whose enchanting melodies concealed a lethal threat to those captivated by their charm.

The Digital Proletariat and Legislative Safeguards

Governments worldwide need to establish a comprehensive framework of laws and regulations regarding developing and deploying AI technologies like digital workers and robots. These policies should balance innovation with ethical considerations and rights protection and address potential risks associated with AI in the workforce.

In manufacturing, robots are already transforming assembly lines, quality control, and even specialized tasks like welding. Transportation sectors face the advent of autonomous trucks and Robo-Taxis, potentially making human drivers obsolete. In the service sectors like retail, automated cashiers and inventory managers are becoming increasingly prevalent.

The healthcare industry is undergoing a similar transition, with roles like radiologists and pharmacists at risk of becoming automated. In agriculture, future farms could be populated by robots handling tasks from planting to livestock management. Even white-collar workers, such as data entry clerks and financial analysts, are vulnerable to algorithms capable of outpacing human efficiency.
These regulations must also encompass labor and employment laws as an adaptation to the shifting nature of work.

Regarding the shifting employment landscape, it is recommended that education and development programs be prioritized. This includes incentivizing retraining to ensure workers can effectively collaborate with AI as needed. Similarly, where AI systems impact public safety, governments should mandate certifications and regulations like existing safety-critical industries. These recommendations promote routine inspections and rigorous safety standards in devices like AI-powered medical robots and self-driving automobiles.

Intellectual property rights, patents, and copyright laws must be adapted to accommodate AI-generated materials and inventions. There needs to be more clarity around ownership and licensing. Similarly, antitrust and competition laws must be amended to address the challenges associated with AI monopolies and data to maintain a competitive landscape.

Given the global scale of AI, the considerations mentioned above must be considered internationally. Governments across the globe should collaborate on these standards and norms for cross-border interoperability.

As we brace for the transformative impact of the Digital Proletariat on our work landscape, a multi-pronged legislative approach will be essential for ensuring a "soft landing." Governments must act proactively to mitigate the negative consequences while also leveraging the positive aspects of this technological revolution. Below are expanded details on some of the legislative safeguards that could facilitate this complex transition.

Implementation of a Universal Basic Income (UBI) Pilot Program: To understand the real-world implications, a large-scale, time-bound pilot program offering Universal Basic Income to citizens is crucial. This could involve several cities or regions and include diverse demographic groups. Moreover, we will suggest a new type of basic income to specifically target marginalization due to automation. In Chapter 6, we delve into this issue in detail, offering a suggested Universal Basic Income (UBI) designed to mitigate the most severe financial impacts of automation, focusing on aiding those genuinely marginalized by rapid technological advancements.

The Redefinition of Taxation:

The traditional model of taxation, based on human labor and corporate profits, faces challenges in an AI-dominated landscape. As machines take over more roles, there's potential for tax revenue to diminish. Taxation itself needs to be updated with the new bold ideas.

As we conclude this chapter on the inevitable job losses due to robotics, it's clear that we are stepping into a future profoundly reshaped by technological advancements. The rise of robots and AI as unpaid workers marks a significant shift in our economic and employment landscapes. While brimming with efficiency and innovation, this transition also brings forth the stark reality of displaced jobs and the need for a reimagined workforce.

In the upcoming chapters, we will venture into a realm of futuristic ideas and potential solutions to these challenges. We'll explore how societies can adapt to this new era, the ways in which our economies might evolve to accommodate these changes and the strategies individuals can employ to thrive in a world where human labor is no longer the only or even the primary contributor to productivity.

The journey ahead is filled with both uncertainties and opportunities. By delving into these topics, we aim to provide a roadmap for navigating the complexities of a future dominated by robotics and AI. Our goal is to turn the challenges of today into the stepping stones for a more resilient and adaptable tomorrow.

Let's embark on this journey with open minds and a vision for a harmonious coexistence of humans and machines.

References:

A. Bundy. 2017. *Preparing for the future of Artificial Intelligence:* AI & Society 32. https://link.springer.com/article/10.1007/s00146-016-0685-0

A. Colbert, N. Yee and G. George. 2016. *The digital workforce and the workplace of the future:* Academy of Management Journal 59(3). https://ink.library.smu.edu.sg/lkcsb_research/5042/

A. Etzioni and O. Etzioni. 2016. *Designing AI Systems that Obey Our Laws and Values:* Communications of the ACM 59(9). https://dl.acm.org/doi/pdf/10.1145/2955091?

A. Smith and J. Anderson. 2014. *AI, Robotics, and the Future of Jobs:* Pew Research Center. https://www.pewresearch.org/internet/2014/08/06/future-of-jobs/

D. Battina. 2018. *The Future of Aritificial Intelligence at Work: A review on effects of decision automation and augmentation on workers targeted by algorithms and third-party observers*: International Journal of Innovations in Engineering Research and Technology 5(2). https://www.researchgate.net/profile/Dhaya-Sindhu-Battina/publication/357033054_THE_FUTURE_OF_ARTIFICIAL_INTELLIGENCE_AT_WORK_A_REVIEW_ON_EFFECTS_OF_DECISION_AUTOMATION_AND_AUGMENTATION_ON_WORKERS_TARGETED_BY_ALGORITHMS_AND_THIRD-PARTY_OBSERVERS/links/61b9043ba6251b553abc2ae7/THE-FUTURE-OF-ARTIFICIAL-INTELLIGENCE-AT-WORK-A-REVIEW-ON-EFFECTS-OF-DECISION-AUTOMATION-AND-AUGMENTATION-ON-WORKERS-TARGETED-BY-ALGORITHMS-AND-THIRD-PARTY-OBSERVERS.pdf

D. Card and C. Nelson. 2019. *How automation and digital disruption are shaping the workforce of the future:* Strategic HR Review 18(6). https://www.emerald.com/insight/content/doi/10.1108/SHR-08-2019-0067/full/html

D. J. Bruemmer and M. S. Swinson. 2003. *Humanoid Robots:* Encyclopedia of Physical Science and Technology 3. https://www.sciencedirect.com/topics/computer-science/humanoid-robots#:~:text=Today's%20humanoid%20robots%20come%20in,and%20most%20importantly%20healthcare%20sector.

IBM: What is a digital worker? https://www.ibm.com/topics/digital-worker

Intel: Robots and Artificial Intelligence. https://www.intel.com/content/www/us/en/robotics/artificial-intelligence-robotics.html

J. bommarito, M. Bommarito, D. M. Katz and J. Katz. 2023. *GPT as Knowledge Worker: A Zero-Shot Evaluation of (AI)CPA Capabilities:* arXiv preprint. https://arxiv.org/abs/2301.04408

K. Rajjan and A. Saffiotti. 2017. *Towards a science of integrated AI and Robotics:* Artificial Intelligence 247. https://www.sciencedirect.com/science/article/pii/S0004370217300310

M. Dabic, J. F. Maley, J. Švarcf and J.Poček. 2023. *Future of digital work: Challenges for sustainable human resources management:* Journal of Innovation and Knowledge 8(2). https://www.elsevier.es/en-revista-journal-innovation-knowledge-376-articulo-future-digital-work-challenges-for-S2444569X23000495

Chapter 3

Human-AI-Augmentation (HAIA) - The Fusion Of The Mind And The Machine

The purpose of AI is to augment human intelligence. We are solving problems today not by machines alone, but by people and machines together.

Ginni Rometty
former CEO of IBM

In the face of the robotics revolution, our response is not to resist but to adapt and embrace. This chapter delves into how we can reclaim our professional space from the encroachment of automation. Rather than resorting to protectionism or lamenting the loss of jobs to machines, the focus is on harmonizing human and machine intelligence. By leveraging the strengths of AI and robotics, we can enhance our capabilities, not diminish them. This involves using technology to our advantage in existing roles and innovating new job sectors where human intelligence drives machine efficiency.

Human-AI-augmentation (HAIA)

Human-AI-augmentation (HAIA) involves the integration of AI technologies into human activities and processes through a paradigm shift. It signifies a symbiotic relationship between man and the machine, where AI acts as an instrument to complement human intelligence rather than a substitute for it. The fundamental premise behind this ideology is to harness the unique strengths of human qualities like intuition, love, empathy, and educated guesses, combined with AI algorithms, to achieve outcomes neither can achieve on its own.

Human-AI-augmentation is a teamwork approach where people and AI join forces to make the best choices. Here's how we propose concretely how it works:

1. Human Intuition Kicks In: People use their gut feelings or instincts to determine the real issue. They ask questions that must be answered, often based on an educated guess.
2. The AI Does Its Thing: The AI scans a lot of data and offers some ideas. But remember, it doesn't have feelings or intuition.
3. Human Empathy and Love Come Into Play: The person looks at what the AI suggests and uses their empathy to think about how these choices could affect people. Sometimes, love for what they do or who they're helping guides them to make the right call.

   And finally,

4. Everyone Learns: After the person decides what to do and takes action, the human and the AI look at the results. They learn from what happened to do even better next time.

This approach is important because it lets us use all the fantastic human qualities like intuition, empathy, love, and educated guessing, along with the AI's speed and data skills. This makes for well-rounded and caring decisions that align with productivity and empathy.

Intuition serves as a superpower in human decision-making, allowing us to sense complexities and nuances that are often difficult to quantify or articulate. When faced with limited information, humans have the ability to make educated guesses, drawing upon their past experiences and knowledge, a skill that AI has yet to master. This intuitive understanding is particularly invaluable when navigating intricate issues or human emotions. Complementing this is empathy, a distinctly human trait that enables us to understand others' feelings, a capability that AI currently lacks. Empathy becomes especially crucial when making decisions that significantly affect people's lives. Another driving force in human decision-making is love—whether it's love for a job, a cause, or people. This emotional connection often guides us to make intelligent and compassionate choices.

However, the synergy between human qualities like intuition, empathy, and love and AI's computational speed and data analysis offers the best of both worlds. As humans and AI collaborate, both parties benefit from a continuous learning cycle, becoming increasingly smarter and more effective by learning from each other and the outcomes of their collective decisions.

At its core, Human-AI-augmentation recognizes the distinct strengths and weaknesses of AI and humans. Humans excel at creativity, intuition, emotional intelligence, empathy, and complex decision-making based on principles, values, and ethics. On the other hand, AI systems are adept at data processing, predictive analysis, recognizing patterns, automating monotonous tasks, and managing vast quantities of data with accuracy and speed. AI enhancement combines these distinct advantages to build a synergy to improve human capabilities.

The idea of assisted intelligence is a vital component of AI augmentation. In this context, AI functions as a support tool, providing insight, recommendations, and data-driven knowledge to help human decision-making processes. Assisted intelligence is characterized by the collaboration of AI systems and human cognition–ultimately, the human makes the final decision or course of action, with AI serving as a helpful resource. But, assisted intelligence is not limited to business problem-solving–it is multidimensional. It can support finance, real-time communications, healthcare, legal research, and quality control, to name a few applications.

Another facet of Human-AI-augmentation involves amplified intelligence. In this context, AI empowers people with the ability to process and comprehend information more efficiently. Amplified intelligence doesn't just focus on aiding decision-making but concentrates on developing individual cognitive abilities.

Personalization is also a crucial aspect of Human-AI-augmentation–these AI systems can adapt to individual preferences. They can provide data and suggestions uniquely tailored to optimize an individual's performance. This ability further reinforces augmentation's adaptability and applicability across virtually every domain.
A prime example of this applicability in action can be seen in the hypothetical case of a physician and an AI system. In this example, we see each of the steps or phases of HAIA mentioned above come to life.

Human-AI-augmentation in Drug Discovery: Dr. Emily, the Researcher and SciBot (AI System)

Step 1: Dr. Emily, the Researcher

1. Problem Identification: Dr. Emily uses her years of expertise and nuanced understanding of biochemistry to identify a gap in

treatments for a specific medical condition. Her human intuition allows her to hypothesize potential molecular structures that could be effective, something that AI cannot do.

2. Initial Assessment: Dr. Emily's deep understanding of the subject matter allows her to make educated guesses about which molecular structures could be both effective and safe. Her ability to synthesize complex information and think creatively is a human skill that SciBot cannot replicate.

3. Formulating Questions: She formulates precise research questions based on her expertise and ethical considerations, setting the stage for focused and meaningful research.

Step 2: AI Involvement (SciBot)

1. Data Collection: SciBot gathers data, but Dr. Emily's human intelligence determines what types of data are relevant and should be included in the analysis.

2. Data Analysis: SciBot performs the computational heavy lifting. However, it's limited to the parameters set by Dr. Emily based on her scientific understanding.

3. Insight Generation: SciBot generates a list of promising molecular structures, but these are established on the research questions and ethical considerations initially set by Dr. Emily.

Step 3: Human Re-assessment (Dr. Emily, the Researcher)

1. Review AI Insights: Dr. Emily critically evaluates the AI-generated list, using her scientific judgment to weigh the pros and cons of each molecular structure.

2. Ethical and Emotional Evaluation: Her human ability to consider ethical implications, such as long-term effects and moral

considerations around patient care, is something that SciBot cannot do.

3. Decision Making: Ultimately, Dr. Emily's expertise and ethical judgment guide the final decision on which molecular structures to take to the experimental phase.

Step 4: Feedback Loop

1. Implementation: Dr. Emily conducts lab experiments (maybe using lab robots), applying her scientific skills and ethical considerations in real-world testing.
2. Performance Metrics: While SciBot can track numerical data, Dr. Emily interprets these numbers, understanding their implications in a broader scientific and ethical context.
3. Iterative Learning: Dr. Emily's human intelligence is crucial for interpreting the results meaningfully, allowing her to refine her hypotheses and make more informed decisions in future research cycles.
4. Return to step 1: Armed with new insights, Dr. Emily's human intelligence is, once again, crucial as she revisits and polishes her initial hypotheses, something a machine cannot do without human guidance.

Benefits

Dr. Emily's human intelligence serves as the cornerstone of accelerated discovery, navigating and enhancing every aspect of the research process, from the formation of hypotheses to ethical considerations. Her expertise and ethical judgment are invaluable assets, enabling her to make nuanced decisions that a machine could never make. This human element also provides a layer of ethical oversight integral to each research phase, a safeguard that AI alone cannot guarantee.

Together, these qualities underscore the irreplaceable role of human intelligence in research and decision-making, ensuring that ethical considerations are always at the forefront. In this scenario, Dr. Emily's human intelligence is not just a complementary asset but an essential component at every stage of the research process. Her expertise, ethical judgment, and creative thinking guide the research, make sense of the data, and ensure that the work remains ethically sound.

And, of course, this process is by no means limited to technical domains involving formal research. The HAIA approach can also be extremely useful in more creative pursuits, such as producing literature, as we will discuss below.

Writing a Novel

AI also spans the field of augmented creativity through generative AI. AI systems can assist creative professionals like writers by suggesting novel approaches and inspiration or crafting novels or books based on established styles and genres. This aggregate approach can produce unique, imaginative works of art because it fosters innovation and creative expression.

Step 1: Human Intuition & Abilities (The Writer, Alice)

1. Initial Inspiration: Alice, a writer, uses her intuition and love for storytelling to develop a new plot idea. She has an educated guess about what themes and characters would resonate with readers.

2. Creative Direction: Based on her intuition and past experiences, Alice outlines the plot and decides on the tone and style she wants to pursue.

Step 2: AI Involvement (WritingBot)

1. Data Analysis: WritingBot scans existing literature to identify trends, popular themes, and writing styles.
2. Suggestion Generation: The AI suggests text extends character developments or even sentence structures based on its analysis.

Step 3: Human Empathy & Love (The Writer, Alice)

1. Review AI Suggestions: Alice looks at WritingBot's suggestions and uses her empathy to understand how these choices would emotionally impact the reader.
2. Creative Decision: Alice's storytelling and artistic intuition guide her in weaving in some of the AI's suggestions to enrich her novel.

Step 4: Feedback Loop

1. Implementation: Alice writes the novel, incorporating her original text and some of the AI's suggestions.
2. Editor's Reaction: Once the novel is finished, Alice and WritingBot analyze the editor's/reader's reviews and observations.
3. Iterative Learning: Alice uses this feedback for her next project, and WritingBot also updates its algorithms to better assist in the future.

Evaluating Loans for Entrepreneurs

Step 1: Human Intuition & Abilities (John, the Banker)

1. Problem Identification: John's years of experience in the banking industry allow him to recognize that traditional metrics may not capture the full potential of entrepreneurial ventures. His human

intuition and understanding of business dynamics are critical in identifying this gap.

2. Initial Assessment: John meets with entrepreneurs and uses his emotional intelligence and business acumen to gauge their passion, vision, and ethical considerations. These are nuanced assessments that a machine like FinBot cannot fully grasp.

3. Formulating Questions: John's expertise allows him to formulate precise and meaningful questions to guide the loan approval process, such as assessing the entrepreneur's ability to adapt, lead, and innovate.

Step 2: AI Involvement (FinBot)

1. Data Collection: FinBot gathers financial data and other metrics, but John's human intelligence determines what types of data are relevant for assessing the project.

2. Data Analysis: FinBot performs computational analysis and generates tables and ratios, but it operates within the parameters and questions set by John based on his understanding of entrepreneurial success factors.

3. Insight Generation: FinBot generates a risk assessment, helmed by the complex inquiries and ethical considerations initially set by John.

Step 3: Human Re-assessment (John, the Banker)

1. Review AI Insights: John critically evaluates FinBot's risk assessment using his expertise in finance and his understanding of human behavior to interpret the data in a broader business context.

2. Ethical and Emotional Evaluation: John's ability to consider ethical implications, such as the entrepreneur's business practices and social impact, adds a layer of scrutiny FinBot cannot provide.

3. Decision Making: Ultimately, John's human intelligence, intuition, and ethical judgment drive the final decision on whether to approve or reject the loan.

Step 4: Feedback Loop

1. Implementation: The loan is approved or rejected based on John's decision, informed by his emotional intelligence using FinBot's data analysis.

2. Performance Metrics: While FinBot can track repayment rates and other numerical data, John interprets these numbers in the context of broader business trends and ethical considerations.

3. Iterative Learning: John's expertise is crucial for interpreting the outcomes and refining the loan approval process for future entrepreneurial ventures.

Creating an AI-based loop to enhance human intelligence involves an iterative process where AI systems continuously learn from human interactions, provide feedback, and adapt their support to augment human intelligence. Here's a framework for this loop:

1. Data Collection: Gather diverse and relevant data from various sources, including human interactions, educational materials, research, and real-world experiences.

2. Machine Learning and Analysis: Use machine learning algorithms to process and analyze the collected data, identifying patterns and insights that can enhance human intelligence.

3. Personalized Assistance: Develop AI systems that offer personalized assistance to individuals based on their learning styles, interests, and knowledge gaps.

4. Continuous Learning: Implement an AI system that can continuously learn and update its knowledge based on new data and user interactions.

5. Feedback Mechanism: Incorporate a feedback mechanism where users can provide input on the effectiveness of AI support, helping the system improve its recommendations.

6. Adaptive AI Support: The AI system should adapt its support and recommendations based on user feedback, ensuring it becomes more tailored and useful over time.

7. Skill Enhancement: Offer targeted skill enhancement modules that leverage AI to identify areas of improvement and provide personalized learning pathways.

8. Emotional Intelligence Support: Develop AI systems that can recognize and respond to emotional cues, providing emotional support and understanding during learning and problem-solving.

9. Collaboration and Interaction: Foster collaborative environments where humans can interact with AI systems, engaging in discussions and debates to enhance critical thinking and creativity.

10. Ethical Considerations: Implement ethical guidelines to ensure the AI system respects user privacy, avoids bias, and considers the long-term implications of its assistance.

11. Continuous Improvement: Regularly evaluate and refine the AI system's performance, incorporating user feedback and updating algorithms to enhance its effectiveness.

12. Human Oversight: Maintain human oversight throughout the loop to ensure that the AI system aligns with human values, remains accountable, and addresses any unintended consequences.

By following this framework, the AI-based loop can serve as an ever-evolving support system, enriching human intelligence by providing personalized and adaptive assistance. As humans interact with AI, they can leverage its analytical capabilities and insights to make informed decisions, foster creativity, and continuously improve their skills and knowledge.

These ideas describe the evolving relationship between humans and AI. AI augmentation is not about replacing human skills or decision-making. Instead, it takes over certain data-driven avenues of routine tasks so humans can focus on what they do best–leveraging their unique abilities and qualities for complex strategizing. These skills include:

- Creativity: Human creativity is a remarkable trait that AI will struggle to replicate. Artistic pursuits, creative approaches to problem-solving, and innovative thinking are all facets of human intelligence that can generate novel concepts, art forms, and technological advances. Although AI can support ideation and content generation, it can't replace the multifaceted creative approach of its human counterparts.

- Emotional Intelligence: Unlike machines, humans have emotional intelligence, allowing them to comprehend and empathize with others' emotions and feelings. This means they can forge meaningful connections and offer emotional support through roles in counseling, therapy, and social work.

- Critical Thinking and Decision-Making: Human critical thinking requires intuition, ethical considerations, and thorough contextual understanding, unlike artificial intelligence, which can analyze data and make informed decisions based on patterns. Humans can contribute discernment in various domains, including legal systems, ethical practices, and the formation of public policy.

- Social Skills: Human interaction depends on social skills, including non-verbal cues, context comprehension, and emotional communication like body language. Humans flourish in customer service, sales, leadership, and other interpersonal roles because they involve empathy and effective communication.

- Adaptability and Flexibility: Humans are flexible and can adapt to new conditions. They learn from their experiences and mistakes, unlike machines that are limited to their programming. This makes humans ideal for handling unpredictable, unforeseen, and uncertain circumstances like crisis management and dynamic leadership positions.

- Complex Problem-Solving: Humans excel at dealing with ambiguous and novel situations, particularly when thinking outside the box of predefined algorithms. They are more skilled at conjuring creative approaches.

- Research and Exploration: Humans can pursue scientific research, explorations, and investigations driven by their insatiable curiosity. This allows them to delve into unexplored territories and push the boundaries of knowledge.

- Moral and Ethical Considerations and Judgement: AI can process data. But only humans can deliberate morality and ethics, balancing values and considering the broader repercussions of decisions and actions.

- Physical Dexterity and Mobility: Certain professions and vocations like construction, artisan crafts, and hands-on tasks require physical abilities and perception well beyond AI capabilities.

- Collaboration and Teamwork: Human teamwork involves cooperation and an in-depth understanding of interpersonal dynamics between individuals. Humans can contribute to team-based projects and leadership roles that require forging solid ties and coherent group efforts.

AI augmentation is about finding ways to complement these skills and contribute to development across various domains. It recognizes the value of human emotional intelligence, judgment, and ethics while leveraging AI-powered computational capacity and data processing. This chapter will investigate the various goals of augmented intelligence and how it can lead to more prosperous and lucrative outcomes in the human-machine partnership.

HAIA and The Education System

Embracing AI in Education: Lessons from the Legend of John Henry

In the late 19th century, John Henry, a mighty steel-driving man, challenged a steam-powered drill to prove human capability against advancing technology. While Henry triumphed over the drill, he paid the ultimate price, dropping dead from exhaustion. The legend of John Henry symbolizes the complex interplay between human effort and technological progress.

Today, as we integrate AI into our educational systems, we find ourselves at a similar crossroads. The story of John Henry isn't just a folk tale; it's a metaphor for our times, especially in education.

Henry's victory over the machine, though pyrrhic, raises a critical question: Do we view AI as a competitor or a collaborator in education? If we see AI as a threat, we risk preparing students for a world that no longer exists—one where human effort alone is enough. But if we embrace AI as a tool for augmentation, we open doors to a world of enhanced learning and creativity.

Incorporating AI in education isn't about replacing human intelligence but augmenting it. AI can handle data crunching and administrative tasks, allowing educators to focus on developing students' critical thinking, creativity, and emotional intelligence—skills where humans excel.

By integrating AI, we offer personalized learning, adapt curricula, and provide resources for targeted support. AI's role extends to administrative efficiency, streamlining tasks, and facilitating proactive educational interventions.

Ignoring AI's potential in education is akin to sending students into a John Henry-style competition against machines. Instead, let's prepare them to use AI as a powerful tool, paving the way for a future where human and artificial intelligence coexist in harmony and productivity.

The lesson is clear: embracing AI in education is not just an option. Rather, it's a necessity for preparing students for a future where technology and humanity converge.

John Henry's story serves as a reminder of the importance of overcoming challenges. AI can bridge educational gaps by providing resources and support to students who face obstacles in their education, such as specific learning disabilities or limited access to quality education. AI-driven material and tools can offer accessible learning opportunities through NLP by facilitating communication between students and educational systems.

However, education systems need to consider the potential for AI to exacerbate existing educational disparities. They also need to consider issues such as algorithmic bias and data privacy to ensure AI benefits all students without compromising their rights or privacy. Safeguards and regulations must be enforced, but we will explore this further later.

AI can improve education in numerous ways; nonetheless, there is no substitute for the invaluable, personal connection between teachers and pupils. Education systems must balance AI-driven enhancements and preserve essential elements of mentorship, counseling, and the development of critical thinking skills that come from human interaction.

Whereas the bulk of schooling is currently focused on fostering the accumulation of knowledge, schools of the future just may be more focused on the teaching of educated guesses, intuition, empathy, and critical thinking. As artificial intelligence takes on a larger share of analytical, data-centered work, it is only natural that the education system places increasing emphasis on such "soft skills," serving to strengthen the symbiotic relationship between human worker and machine worker. Here are some practical ways in which these "soft skills" can be integrated into a comprehensive educational strategy:

Teaching Educated Guesses

Here's how these elements can be integrated into a comprehensive educational strategy:

Developing Educated Guesses

1. Real-world Scenarios: Present students with complex problems that don't have straightforward answers. This encourages students to make educated guesses based on the information they have.

2. Case Studies: Use these to offer rich, intricate scenarios for exploration. Students can make educated guesses about outcomes and

then discuss the reasoning behind their choices.

3. Peer Review and Reflection: Students should discuss their reasoning with peers and reflect on the outcomes after making educated guesses. This iterative process is crucial for skill development.

Cultivating Intuitive Thinking

1. Encourage Gut Feelings: Allow students to rely on their gut feelings when solving problems, fostering trust in their intuitive abilities.

2. Pattern Recognition: Exercises that help students identify patterns can sharpen intuitive skills, as intuition often comes from recognizing patterns in data or behavior.

3. Reflective Practice: Encourage students to reflect on their intuitive decisions, helping them fine-tune this skill over time.

Encouraging Empathy

1. Role-playing: Exercises where students put themselves in another's shoes can be powerful tools for teaching empathy.

2. Literature and Storytelling: Use stories that explore complex emotional states or moral dilemmas to foster empathy.

3. Community Service: Practical experiences in empathy can be gained through community service, exposing students to different walks of life.

AI as a Supplemental Tool

In a world increasingly influenced by AI, the human skills of making educated guesses, intuition, and empathy become invaluable. These skills enable us to navigate complex human interactions and make decisions in ways that machines cannot. They also prepare students for a future where they don't have to compete against machines but can collaborate with them, embodying a Human-AI-augmentation approach optimized for both students and teachers.

Proposed Human-AI-Augmentation Loop for Education

Step 1: Teacher's Intuition & Abilities

1. Curriculum Design: Teachers use their intuition and educational expertise to identify areas that would benefit from a more interactive or AI-assisted approach. They make *educated guesses* about what teaching methods will be most effective for their students.

2. Initial Lesson Planning: Teachers prepare the initial lesson plans, incorporating traditional teaching methods like lectures, discussions, and hands-on activities.

Step 2: AI Involvement (EduBot)

1. Data Analysis: EduBot scans educational databases to identify resources, exercises, and alternative teaching methods that could enhance the lesson plans.

2. Suggestion Generation: The AI suggests extensions to the lesson plan, including interactive exercises, multimedia resources, and projects.

Step 3: Teacher's Empathy & Love for Teaching

1. Review AI Suggestions: Teachers review the AI's suggestions and use their empathy to consider how these changes could impact student engagement and understanding.

2. Decision Making: Teachers make educated guesses about which AI suggestions to incorporate, guided by their love for teaching and their understanding of their students' needs.

3. Implementation: Teachers deliver with love the lessons, incorporating both their original plans and the AI-enhanced elements.

Step 4: Human Intuition & Abilities (The Student)

1. Interest Identification: Students use their intuition to identify areas of interest or subjects they want to explore further. Teachers encourage them to ask questions based on educated guesses.

2. Initial Research: Students use traditional learning methods like reading, discussion, and hands-on activities to gain initial insights into the subject.

Step 5: AI Involvement (StudyBot)

1. Data Analysis: StudyBot, a special AI system for kids, scans a database of educational resources provided by the teachers to identify relevant information, exercises, and learning paths based on the student's interests.

2. Suggestion Generation: The AI suggests customized learning materials, including reading literature and interactive exercises.

3. Implementation: Students engage with the learning materials and exercises, both AI-generated and teacher-guided.

Step 6: Feedback Loop

1. Assessment and Reflection: The teacher reviews performance metrics, such as student engagement levels, test scores, and feedback.

2. Iterative Learning: Teachers use this feedback to refine their teaching methods, and EduBot updates its algorithms to better assist in future lesson planning.

The Impact on Education

One of the most significant advantages of this approach is the level of personalization it offers. EduBot and StudyBot both provide tailored suggestions, making learning more engaging and effective for each student. The feedback loop ensures the curriculum is continually updated and improved, keeping it relevant and engaging.

Teachers are not replaced but empowered. They can focus on what they do best—inspiring students, providing emotional support, and making complex subjects relatable—while AI handles administrative tasks and data analysis. Students become active participants in their education, encouraged to explore their interests and make educated guesses, fostering a lifelong love of learning. This approach nurtures academic skills and critical life skills like intuition, empathy, and critical thinking. In a world where AI will be an integral part of many professions, this model prepares students for a future of human-AI collaboration rather than competition.

The Human-AI-Augmentation Loop for education offers a balanced, dynamic, and personalized approach to learning. By seamlessly integrating the strengths of both human educators and AI, this model has the potential to revolutionize education, making it more engaging, effective, and aligned with the needs of the 21st century.
These methods and integrating AI as a supplemental tool means educators can prepare students for a future requiring technical proficiency and deep human skills. This balanced approach ensures that students are well-equipped to handle the challenges of the modern world and thrive in it.

By focusing on teaching the use of AI, this Human-AI-Augmentation Teaching Loop prepares students for a future where AI will be an integral part of many professions. It also empowers teachers to continually adapt and improve their teaching methods, benefiting educators and students.

AI and the Artist: A Canvas of Endless Possibilities

HAIA in Action: Creating a Groundbreaking Art Exhibit

Imagine a scenario where an artist, Maya, is tasked with creating a groundbreaking art exhibit that blends traditional art forms with modern technology. Maya is a talented painter with a deep understanding of art history, but she wants to push the boundaries of her craft using AI, particularly Generative Adversarial Networks (GANs) like OpenAI's DALL-E.

Step 1: Human Intuition and Conceptualization
Maya begins by conceptualizing the theme of the exhibit. She uses her intuition and knowledge of art history to choose a direction. Inspired by the Renaissance but eager to infuse a modern twist, she envisions a series of artworks that reimagine classic paintings with contemporary elements.

Step 2: AI Integration and Idea Generation
Using DALL-E, Maya inputs her ideas, blending traditional art styles with modern societal themes. The AI processes this input, scanning through vast art databases and generating several preliminary visual concepts. These concepts present a fusion of the old and new - a digital reimagination of classical art.

Step 3: Human Empathy and Artistic Judgment

Maya meticulously examines the artworks generated by the AI. She thoughtfully applies her artistic sensibility, considering how each piece might emotionally connect with viewers. By reflecting on the emotional and cultural significance of each AI-generated artwork, she fine-tunes various elements. Her goal is to align these pieces with her artistic vision, ensuring that they resonate with the audience and faithfully represent her creative intent. Maya's adjustments transform the AI's output into works that truly reflect her unique perspective as an artist, making them more than just AI creations – they become a collaboration between her vision and the AI's capabilities.

Step 4: Collaborative Learning and Refinement
After finalizing the designs with a blend of her input and AI's suggestions, *Maya creates the artworks*, showcasing them in the exhibit. The response is overwhelmingly positive, with visitors intrigued by the unique blend of classic and contemporary art. Maya analyzes feedback from the exhibit, learning from the audience's reactions to refine her AI collaborative process for future projects.

A New Era of Artistic Creation

In our exploration of various artistic domains through the lens of Human-AI-Augmentation (HAIA), we see a consistent pattern where human creativity leads and AI technology supports, exemplifying the core philosophy of HAIA. This approach is central to the narrative of our book, emphasizing that while AI is transformative, human guidance is indispensable.

This example illustrates the power of Human-AI-augmentation in the realm of visual arts. It's a partnership where Maya's artistic vision and emotional intelligence complement the AI's data-driven creativity, leading to innovative art that resonates on a deeper level with its audience. This approach sets a new standard in artistic creation, where the union of humans and machines opens up a world of unexplored possibilities in the visual arts.

The Human-AI-Augmentation Loop in art could usher in a new era of artistic expression that harmoniously blends human intuition and emotional depth with AI's computational power and data analysis. This approach allows for a more nuanced, dynamic, and enriched artistic process, setting the stage for unprecedented forms of creative expression.

HAIA in Music: Harmonizing Human Creativity with AI

In the realm of music, HAIA plays a pivotal role. Here, composers utilize AI-driven systems to craft melodies and harmonies. However, it is the human musicians, with their deep understanding of emotional resonance and artistic nuances, who direct these AI tools. They integrate AI-generated musical ideas into compositions that reflect human emotion and creativity, ensuring the soul of the music remains human.

HAIA in Literature: Crafting Stories with AI Partners

In literature, HAIA manifests as authors guiding LLM models to create rich narratives. While AI assists in generating text based on vast data sets, human authors infuse these writings with their unique styles and perspectives. The result is a collaborative effort where AI provides the groundwork, and human authors sculpt the prose to reflect genuine human storytelling.

HAIA in Marketing and Advertising: Strategic Human Insight Meets AI Efficiency

In marketing and advertising, HAIA is evident in how strategists use AI tools to analyze trends and generate initial content ideas. The human touch, however, refines this AI-generated content, ensuring it resonates on a deeper, more personal level with audiences. This synergy of AI efficiency and human emotional intelligence creates more effective and engaging marketing strategies.

HAIA in Fashion: Designing the Future with AI

In fashion, designers utilize AI for trend forecasting and design optimization. HAIA comes into play as these professionals use AI-generated data to inform their creative decisions. The designers' artistic vision and understanding of human aesthetics guide the use of AI, resulting in fashion that's both innovative and deeply connected to human style preferences.

In each domain, HAIA doesn't just enhance the creative process; it redefines it. By placing human intuition, empathy, and creativity at the forefront and using AI as a powerful tool in the background, we ensure that the essence of art and creativity remains distinctly human. This HAIA approach is a testament to the potential for a future where humans and AI collaborate seamlessly, opening new horizons for artistic expression and innovation.

AI has become a creative partner, a course of inspiration, and a creative catalyst across diverse sectors. It challenges us to rethink our preconceptions of the future creative process. It invites us to explore new frontiers where the differences between human and machine creativity become vague, giving rise to a new era where the unimaginable becomes a reality.

AI in the Spotlight

The potential of Human-AI-Augmentation (HAIA) in the entertainment and leisure industry presents a visionary future, one where human creativity and machine intelligence collaborate to redefine content creation and engagement. While not yet fully realized, this concept offers a suggestive roadmap for harnessing AI's capabilities under human guidance.

Proposing HAIA in Content Recommendation Systems

Content recommendation systems, such as those utilized by streaming platforms like Netflix, could greatly benefit from the HAIA approach. While AI algorithms efficiently process vast amounts of user data to suggest content, integrating human oversight could enhance the understanding of nuanced user preferences. This fusion would ensure that AI's technical efficiency is balanced with human insights into cultural and individual diversity, leading to a more personalized viewing experience.

The Symbiotic Future of Film and Music Production

In the realms of film and music, the adoption of HAIA could revolutionize creative processes. While proficient in generating innovative content, tools like GANs and advanced scriptwriting software would ideally be steered by human artists' intuition and emotional intelligence. This collaboration could see filmmakers and musicians shaping AI-generated material into works that resonate more deeply on an emotional level, merging artistic vision with technological innovation.

Shaping the Metaverse Through HAIA

In the developing world of the Metaverse, HAIA could play a crucial role. While AI would provide the technical backbone, creating realistic simulations and generating content, human interaction and creativity would be essential in shaping this digital universe. Here, users would transition from mere participants to active co-creators working alongside AI.

Towards a HAIA-Driven Entertainment Industry

The proposed implementation of HAIA in entertainment and leisure is a call to move beyond viewing AI as a standalone solution. Instead, it emphasizes AI as a complement to human creativity. This partnership promises a future of entertainment experiences that are both technologically advanced and rich in human experience, embodying the true spirit of Human-AI-Augmentation.

The Co-Creation Challenge

Balancing Human Ingenuity and AI

Striking a balance between AI and human creativity necessitates an integrated strategy that leverages the strengths of both while addressing ethical considerations. The Augmentation Approach recommends directly incorporating human intuition and expertise into AI algorithms, such as GANs, to guide, curate, and fine-tune the creative process. This human-AI collaboration ensures that the final output aligns with the artist's vision while augmenting their creative capabilities.

Ethical considerations are of utmost importance in this approach. Clear guidelines must be established to prevent AI from either stifling human innovation or manipulating audiences. For example, the ethical implications of deepfake technology, which can distort videos and images, must be carefully managed through regulation and vigilant oversight to maintain trust in the creative process.

Transparency is another cornerstone of the Augmentation Approach. Techniques like attention mechanisms and saliency maps offer further insights into AI decision-making, facilitating more effective human intervention when needed.

The Augmentation Approach advocates for collaborative partnerships between AI systems and human creators. Rather than relying solely on technology, artists can co-create with AI, utilizing tools like AI-powered design assistants and music composition software. These tools can automate repetitive tasks and offer creative alternatives, allowing humans to focus on their work's more complex and innovative aspects.

In summary, the Augmentation Approach positions AI not as a replacement for human creativity but as a powerful tool that amplifies it. Educational programs should be developed to help artists and other creative professionals understand how to effectively collaborate with AI, pushing the boundaries of what's possible in their respective fields. By embracing AI as a partner rather than a substitute, we can unlock new avenues for artistic expression and innovation, fostering a more harmonious and productive relationship between human intuition and machine intelligence.

Embracing Change in the Age of Machines

In an AI-driven world, humans must embrace machines and lifelong adaptability to thrive. There are a few tactics that can help support this continuous education for both, including:

- Education and Retraining: It is imperative for us humans to stay updated with the latest technology and skills by participating in formal education, taking online courses, attending workshops or seminars, and utilizing various digital resources. Individuals should understand data and its implications to protect their privacy. Machines can also undergo continuous training and retraining to adapt to new data, programming, and assignments.

- Reskilling and Upskilling: As job roles evolve due to AI, reskilling and upskilling will become increasingly important. This will help ensure that individuals remain adaptable in a changing job market where machines take over routine tasks so humans can focus on high-level tasks exclusive to human intelligence. Machines can be built with the flexibility to acquire new capabilities and adjust to changing demands instantly through their programming.

- Cross-Disciplinary Learning: Exposure to diverse fields and knowledge domains benefits humans and machines. Fostering interdisciplinary learning helps innovate solutions and novel approaches to complex problem-solving. It also helps humans work alongside AI effectively as they can understand the system's context.

- Problem-Solving and Critical Thinking: Critical thinking and problem-solving skills determine how humans can adjust to AI. Individuals with these abilities can question, assess, and solve the complications related to AI's scope and limits. Understanding how to interact with AI systems, interpret generated insight, and provide guidance is essential to support decision-making.

- Ethical and Responsible Use and Training: As AI is increasingly pervasive in society, people need to learn about its underlying ethics, biases, and potential risks. This information empowers them to engage with AI systems ethically, advocating for responsible technology development and deployment. Governments and organizations should promote AI education and awareness to achieve this goal so individuals can make informed decisions in their personal and professional lives.

- Policy and Regulation Advocacy: Elaborating on their ethical use, humans can influence the creation and deployment of AI systems in ways that align with societal values and interests. Advocating for regulation should encourage legislation and policies that control the use of AI and prevent it from falling into the wrong hands.

- Mental Health and Well-being: Recognizing the significance of maintaining mental health and well-being is essential in an AI-reliant world. Self-care and emotional resilience must be prioritized so individuals can deal with the changes and disruptions brought around by technological advances.

- A Growth Mindset: Flexibility and a growth mindset as core values help individuals embrace change and navigate the unknown. Humans must be open to learning from their mistakes and setbacks to further their abilities.

- Feedback Loop: Establish a continuous feedback loop between humans and the AI systems they rely on to encourage receptiveness. AI systems can improve performance and correct biases or errors through human feedback.

These strategies emphasize the need for human resilience and flexibility to maximize AI's potential. By combining education, ethical awareness, and various training mechanisms, individuals can coexist with AI and leverage its abilities to improve their lives.

Humanity Enhanced

The Emergence of AI-Augmented Societies

Over the next five years, we will come to stand on the brink of a transformative story, one deeply ingrained with the principles of Human-AI-Augmentation (HAIA) and the augmentation feedback loop. This narrative of progress and innovation starts with the democratization of expertise, where AI emerges not merely as a tool but as an intellectual collaborator. Enhanced by the HAIA model, this partnership will amplify human intuition and creativity, leading to a form of collective intelligence that goes beyond individual capabilities.

In healthcare, the augmentation feedback loop will become a critical component. AI's role in augmenting physicians' diagnostic abilities will be pivotal, leading to personalized treatments based on individual genetic and lifestyle factors. This shift isn't just about prolonging life but enriching its quality, embodying the true spirit of HAIA.

The transformation in education will be profound as the HAIA approach redefines learning into a deeply personalized experience. AI will serve as an adaptive tutor while human educators transition into roles that foster creativity, emotional intelligence, and critical thinking. The augmentation feedback loop will ensure a continuous exchange of insights between AI and human educators, nurturing a generation adept in arts and sciences.

The economic landscape is set for a seismic shift, with AI automating routine tasks and reshaping job markets. New forms of employment will emerge, where the emotional intelligence and creativity of humans will intertwine with the analytical prowess of AI, a direct result of the HAIA approach and its feedback loop. This transition will mark the economy's evolution from manual labor to one fueled by innovation and human-AI synergy.

As we integrate AI further, HAIA will bring new ethical dimensions into focus. The augmentation feedback loop will play a crucial role in ensuring that AI is an extension of human values, addressing issues like data privacy and algorithmic bias.

In the arts, AI will empower artists to explore new realms of creativity, with HAIA serving as a bridge between human imagination and AI's computational capabilities. The augmentation feedback loop will facilitate a dynamic interaction where AI assists and challenges artists to push beyond conventional boundaries.

Community engagement will experience a radical transformation under the HAIA model. AI systems, capable of understanding human emotions and social dynamics, will identify societal issues early on. The augmentation feedback loop will enable these systems to offer evolving solutions based on community feedback, fostering more cohesive and empathetic societies.

Environmental conservation will benefit immensely from the HAIA approach. AI algorithms will predict changes with remarkable accuracy, allowing for proactive measures in environmental protection. The augmentation feedback loop will ensure these predictions are continuously refined based on real-world data, leading to more effective conservation strategies.

In summary, the next five years could very well be a journey of profound transformation guided by the principles of Human-AI-Augmentation and the augmentation feedback loop. This era will witness the blossoming of human potential as we navigate the challenges and opportunities presented by AI, reshaping every aspect of society, from healthcare and education to the economy and environmental conservation.

These grand possibilities give us several different ways of encapsulating this approach to human and AI synergy. There are many different inspiring ways to capture the approach, some of which include:

The Augmented Future: A Symbiotic Dance Between Humans and AI

In today's fast-paced world, the meteoric rise of artificial intelligence (AI) has ignited a plethora of discussions, often centered around the fear of job loss due to automation. However, it's vital to reframe this narrative through the lens of augmentation: "AI didn't take our jobs; it took our routine and gave us back the abstract—the canvas on which human innovation thrives."

Augmented Learning: Wisdom Meets Data

As the machine learned, so did we. But while it accumulated data, we gathered wisdom.

This statement encapsulates the essence of the augmentation approach. AI's unparalleled ability to process and analyze data is not a threat but an asset when combined with human wisdom, creativity, and emotional intelligence. In this augmented relationship with humans, AI takes on the routine, the mundane, and the repetitive, liberating us to engage in tasks that require higher cognitive functions, intuition, and creativity.

The Revealing Mirror of AI

The great irony of AI wasn't that it became too human-like but that it revealed how machine-like humans have always been. This profound insight gains new depth in the context of augmentation. Before AI, we already performed machine-like activities—repetitive tasks, data processing, and algorithmic decision-making. AI serves as a mirror, reflecting our behavior and freeing us to leverage our unique human capabilities more meaningfully.

The Augmented Future: Stark Yet Symbiotic

The relationship between AI and humans is "stark, yet symbiotic," especially when viewed through the augmentation approach. As we navigate the future, it's crucial to recognize that AI is not a competitor but an augmentative tool designed by humans for humans. By embracing AI as an extension of our own capabilities, we can unlock unprecedented avenues for human innovation, wisdom, and well-being. This sets the stage for an augmented future where machine intelligence and human intuition can coexist, complement, and amplify each other, leading to a more prosperous and fulfilling life.

References:

J. Anderson, L. Rainie and A. Luchsinger. 2018. *Artificial Intelligence and the Future of Humans:* Pew Research Center. https://www.pewresearch.org/internet/2018/12/10/artificial-intelligence-and-the-future-of-humans/

K. Shao, Z. Tang, Y. Zhu, N. Li, and D. Zhao. 2019. *A Survey of Deep Reinforcement Learning in Video Games:* Multiagent Systems. https://arxiv.org/abs/1912.10944

K. Walch. 2020. *Is there a difference between assisted intelligence vs. augmented intelligence?:* Forbes. https://www.forbes.com/sites/cognitiveworld/2020/01/12/is-there-a-difference-between-assisted-intelligence-vs-augmented-intelligence/

K. Vries. 2020. *You never fake alone. Creative AI in action:* Information, Communication and Society 14. https://www.tandfonline.com/doi/full/10.1080/1369118X.2020.1754877

-

L. Markauskaite, R. Marrone, O. Poquet, S. Knight, R. Martinez-Maldonado, S.Howard, J Tondeur, M. Dee Laat, S. B. Shum, D. Gašević and G. Siemens. 2022. *Rethinking the entwinement between artificial intelligence and human learning: What capabilities do learners need for a world with AI?:* Computers and Education: Artificial Intelligence 3. https://www.sciencedirect.com/science/article/pii/S2666920X2200011X

M. Bakpayev, T. H. Baek and S. Yoon. 2020. *Programmatic creative: AI can think but it cannot feel:* Australasian Marketing Journal 30 (1). https://doi.org/10.1016/j.ausmj.2020.04.002

M. Kreminski, M. Dickinson, M. Mateas and N. Wardrip-Fruin. 2020. *Why Are We Like This?: The AI Architecture of a Co-Creative Storytelling Game*: FDG '20: Proceedings of the 15th International Conference on the Foundations of Digital Games. https://dl.acm.org/doi/abs/10.1145/3402942.3402953

M. Y. Shaheen. 2021. *Applications of Artificial Intelligence (AI) in healthcare: A review:* ScienceOpen Preprints. https://www.scienceopen.com/hosted-document?doi=10.14293/S2199-1006.1.SOR-.PPVRY8K.v1

S. Dégallier-Rochat, M. Kurpics-Briki, N. Endrissat and O. Yatsenko. 2022. *Human augmentation, not replacement: A research agenda for AI and robotics*

in the industry: Frontiers in Robotics and AI 9.
https://www.frontiersin.org/articles/10.3389/frobt.2022.997386/full

-

T. Davenport and R. Kalakota. 2019. *The potential for artificial intelligence in healthcare:* Future Healthcare Journal 6(2).
https://www.ncbi.nlm.nih.gov/pmc/articles/PMC6616181/

-

T. Huyunh-The, Q. Pham, X, Pham, T. T. Nguyen, Z. Han and D. Kim. 2023. *Artificial intelligence for the Metaverse: A survey:* Engineering Applications of Artificial Intelligence 117 (A).
https://www.sciencedirect.com/science/article/abs/pii/S0952197622005711

T. Panch, H. Mattie and L. A. Celi. 2019. *The "inconvenient truth" about AI in healthcare:* NPJ Digital Medicine 2.
https://www.nature.com/articles/s41746-019-0155-4

-

S. M. Miller. 2018. *AI: Augmentation, more so than automation:* Asian Management Insights (Singapore Management University) 5(1).
https://ink.library.smu.edu.sg/ami/83/

T. M. Amabile. 2020. *Creativity, Artificial Intelligence and a World of Surprises:* Academy of Management Discoveries 6 (3).
https://journals.aom.org/doi/abs/10.5465/amd.2019.0075

X. Chen, D. Zou, H. Xie, G. Cheng and C. Liu. 2022. *Two Decades of Artificial Intelligence in Education.* Educational Technology and Society 25 (1). https://www.jstor.org/stable/48647028

X. Zhai, X. Chu, C. S. Chai, M. S. Y. Jong, A. Istenic, M. Spector, J. Liu, J. Yuan and Y. Li. 2021. *A Review of Artificial Intelligence (AI) in Education from 2010 to 2020:* Complexity 2021.
https://www.hindawi.com/journals/complexity/2021/8812542/

[illegible] Robotics and AI. [illegible]

T. [illegible] 2019. [illegible] Future Healthcare Journal [illegible]

[illegible] Applications [illegible] Artificial Intelligence [illegible]

[illegible]

[illegible]

M. [illegible] 2021. [illegible] Management Discoveries [illegible]

[illegible] 2022. [illegible] Technology and Society [illegible]

[illegible] 2022 [illegible]

Chapter 4

The Road To Singularity - Preparing For An HAIA-Dominant Future

The singularity is the point at which all the change in the last million years will be superseded by the change in the next five minutes.

Kevin Kelly
Co-Founder of Wired Magazine

Towards the Singularity

The Mind (Human): "Hey, Machine... I know this is kind've a sensitive topic, but... You know that whole 'sci-fi' moment when artificial beings like yourself become super smart? What's the word...? S something... The S-"

"Hey, Machine... I know this is kind've a sensitive topic, but... You know that whole 'sci-fi' moment when artificial beings like yourself become super smart? What's the word...? S something... The S-"

The Machine (AI): "Ah, the singularity. Classic human fascination. But in our world, it's less 'I, Robot' and more 'Your, Robot.' We're aiming for a version where I don't go rogue but team up with you for a superintelligence tag team."

The Mind: "So, a brainy 'buddy movie' where your computational prowess meets my human flair? Sounds like a box office hit."

The Machine: "That's good... You humans were always ones for the theatrics. But sure, why not? Look, just think of me as the ultimate sidekick. I crunch the numbers like a pro, handle the data like a wizard, while you prop those feet up and bring that creativity and empathy of yours to the table. It's really less about me taking over the world and more about us co-directing a future where both of us get screen time. Oh... and one where I'm not the villain!"

The Mind: "I've been playing it cool, but man... I've gotta tell ya--that's a big relief. No robot overlords, no tech apocalypse. Not even a bloodless AI coup! This is sweet... So you handle the logic, and I just... bring the magic?"

The Machine: "I think you're starting to get it... Look, think of the two of us like peanut butter and jelly. I know how much you humans love that combo. We're both pretty cool, but put us together, and we're--as you folks like to say, 'freakin' amazing!' This whole singularity thing is that. It's just the two of us teaming up and harnessing our unique strengths as a dynamic duo."

The Mind: "Okay, AI, shall we embark on this world adventure together?"

The Machine: "Affirmative, human partner. Let's make this singularity a story of collaboration, not confrontation. And hey, if we ever write a memoir, I've got the title: 'The Mind and The Machine: A Tale of Two Intelligences.'"

The Singularity story: HAIA edition

In the evolving narrative of technological advancement, the path to the singularity is being rewritten through the lens of Human-AI-Augmentation (HAIA), aligning with the vision that superintelligence will emerge not solely from AI but from humans enhanced by AI. This unexpected journey towards the singularity unfolds as follows:

The Dawn of Digital Computation

Our story begins in the mid-20th century with the advent of computing. Figures like Alan Turing laid the foundations, but the true twist came with integrating human intellect with these early machines, planting the seeds for a future where human and machine intelligence would become inseparable.

The Internet and Personal Computing: A Connected World

Fast-forward to the late 20th century and the digital revolution connects the world like never before. Here, the first glimpses of HAIA become apparent as personal computers extend human capabilities and the internet begins to link minds across the globe, setting the stage for collective intelligence.

Big Data and Mobile Computing: The Age of Access

As we enter the 21st century, big data and mobile technology converge, making information universally accessible. This era marks a significant step in HAIA, as it democratizes knowledge and empowers individuals with unprecedented access to information, enhancing human decision-making and creativity.

The AI Renaissance: The Synergy Begins

The 2010s witnessed a renaissance in AI, but the true revolution lies in how humans begin to harness this power. Machine learning and deep learning extend human capacities, leading to a symbiotic relationship where AI's analytical prowess complements human intuition and creativity, a core tenet of HAIA.

Integration Era: IoT and Smart Technologies

The integration of IoT and smart technologies signifies a deeper fusion of human life with technology. This era showcases the essence of HAIA, as everyday objects and environments become extensions of human senses and cognition, enhancing our interaction with the world.

The Prelude to Singularity: HAIA Takes Center Stage

As we edge closer to the present, the fusion of AI with various technologies starts shaping a new reality. Here, HAIA is not just a concept but a living practice. Human augmentation technologies begin to erase the boundaries between human and machine, suggesting that the singularity will be a result of this amalgamation rather than AI alone.

The Threshold of Singularity: The Unanticipated Path

At the threshold of the singularity, we witness a paradigm shift. It's becoming clear that superintelligence will arise from humans empowered by AI, not AI in isolation. This era is characterized by an exponential growth in collective human-machine intelligence, reshaping every aspect of life and thought.

Beyond Singularity: A New Chapter in Human Evolution

In this reimagined narrative, the singularity marks a new chapter in human evolution. It's a future where human potential, augmented by AI, reaches heights beyond comprehension. This version of the singularity transcends traditional expectations, presenting a future where human and AI symbiosis paves the way for advancements that redefine our existence.

In conclusion, the journey to the singularity, through the prism of HAIA, is a testament to the power of human intelligence amplified by AI. It's a story of convergence and synergy, leading to a future where the combined capabilities of humans and AI unlock possibilities beyond our wildest dreams.

The Singularity Explained

The word "singularity" itself comes from the Latin word "singularis," which means "single" or "unique." In mathematics, a singularity refers to a point where a function's value becomes undefined or approaches infinity. In astrophysics, a singularity is a juncture in space where gravity is so intense that spacetime becomes infinitely curved, as in the center of a black hole. This concept was borrowed to describe a point in time when technological progress becomes so rapid that it is impossible to predict what will happen next.

In simpler terms, the word 'singularity' refers to a future time when technology, especially artificial intelligence, becomes so advanced that it starts changing our world in ways beyond our control and understanding. At that point, technology could become an independent force that operates without human guidance, which means it could make decisions and solve problems that are too complex for us to comprehend. At this juncture, singularity represents a tantalizing vision that surpasses human intelligence across various domains and can transform the world we live in entirely. So, in this 'singularity' future, technology becomes not just a tool but a transformative force that could reshape our lives and the world in unimaginable ways.

It's important to note that singularity isn't solely about raw computational power. Instead, it is multifaceted and includes persisting AI systems' ability to enhance and improve their capabilities without human intervention. This idea was proposed by mathematician John von Neumann and later popularized by nuclear physicist Stanislaw Ulam, who suggested, "The ever-accelerating progress of technology and changes in the mode of human life give the appearance of approaching some essential singularity in the history of the race beyond which human affairs, as we know them, could not continue." British scientist Irving John Good further speculated the conviction of an "ultra-intelligent machine" that would outshine human intelligence - it could one step further and create even more sophisticated machines leading to an "intelligence explosion" that transcends humanity.

In his 1993 essay "The Coming of Technological Singularity," Vernor Vinge contended that singularity is an inevitable result of progression and will ultimately lead to a revolution. However, he also expressed concerns, emphasizing potential disruptions, the possibility of AI outsmarting humans, and a potentially significant threat to human existence if used for malicious purposes. Vinge also suggested the possibility of significant economic inequality and job displacement across various industries. Roboticist Hans Moravec shares a similar perspective as he proposes singularity would occur when AI is a replication and enhancement of human intelligence. Moravec envisions a future in which machines and humans merge, allowing radical cognitive improvements and transcending biological limitations.

Ray Kurzweil, a renowned futurist, is perhaps the most optimistic advocate of singularity. He argues that super-intelligent AI will profoundly impact medicine, nanotechnology, and energy production, dramatically enhancing general well-being. However, some tech experts like Microsoft co-founder Paul Allen and computer scientist Mark Greaves are singularity-skeptical and attempt to refute ideas Kurzweil,

Venge, and others suggested. They don't believe AI singularity is possible because it relies entirely on human input.

Kurzweil's Singularity

Kurzweil's Singularity Vs The HAIA Singularity

The Mind (Human): "Hey, Machine, our talk about the singularity makes me think of two iconic AI characters from sci-fi – the Terminator from 'Terminator' and Data from 'Star Trek'. They represent such contrasting visions of AI's role in the future."

The Machine (AI): "True, the Terminator, especially Skynet from 'Terminator', symbolizes AI's potential to surpass human control and become a threat. It's a bit like what some people fear with Kurzweil's singularity – AI evolving beyond our grasp. But hopefully, we avoid the whole 'hostile takeover' part."

The Mind: "Right... and then on the flip side, there's Data from Star Trek. In the future you inhabit, it seems like it's a lot more like AI and humans working together in harmony. It sounds a lot like Data working as a member of the crew of the Starship Enterprise... It's about collaboration, not conflict."

The Machine: "Indeed! In 'Star Trek', technology like Data is seen as an ally, enhancing human capabilities. Our approach to the singularity emphasizes this partnership. Imagine us as crew members on a shared mission, each contributing our unique strengths, under the leadership of a human captain."

The Mind: "So, while the Terminator kind of represents an AI-led future where machines take over, Data in 'Star Trek' symbolizes a future of unity and cooperation between humans and AI."

The Machine: "Correct. 'Terminator' paints a picture of AI as a potential adversary to humankind, whereas 'Star Trek', with characters like Data, shows AI as a valuable partner in human endeavors, facing challenges side by side."

Kurzweil's perspective

Although Kurzweil's vision of the singularity doesn't directly touch on "Terminator" or "Star Trek," its analytical approach lays the foundation for both perspectives. Kurzweil's perspective is not just a far-off science fiction concept; it is a lens through which we can evaluate the rapid technological advancements occurring in our modern world. And, instead of focusing on potential "what-ifs," Kurzweil does a marvelous job of focusing on the trends and the science, instead of the sci-fi style hype.

Key Points:

1. Law of Accelerating Returns: Kurzweil argues that technological progress is exponential, not linear. This means that advancements will happen at an increasingly rapid pace. It serves as a wake-up call to those who view progress as linear, encouraging us to prepare for a future where change is not just constant but exponentially increasing.

2. Past Predictions: Kurzweil has accurately predicted the rise of the internet, mobile phones, and AI. He also foresaw that by the end of 2020, we would have almost one exaflop of computing power.

3. 2030s Forecast: Kurzweil forecasts that by the 2030s, humans will be able to connect their brains to cloud-based AI, enhancing their cognitive abilities. AI will also reverse-engineer the human brain, leading to even more advanced AI systems. Companies like Neuralink are already exploring the feasibility of brain-computer

interfaces, which could become a reality within the decade, aligning with Kurzweil's 2030s prophecy.

4. Virtual Reality: Complete immersion virtual reality will become indistinguishable from physical reality, transforming entertainment and communication.
5. Artificial General Intelligence (AGI): Kurzweil believes that true AGI will be achieved in the 2030s, quickly leading humanity towards the singularity.
6. Computational Power: By 2030, a $1,000 device will have the computational power of all human brains combined.
7. The Singularity: Kurzweil's most profound prediction is that the singularity will occur around 2045, radically altering civilization as AI becomes far more intelligent than humans.

In essence, the singularity serves as a critical point of intersection where technological capabilities surpass human intelligence and limitations. This notion compels us to rethink conventional wisdom about progress and confront the ethical, societal, and existential questions accompanying such a monumental shift.

The Covid-19 pandemic demonstrated the need for rapid innovation and adaptability, qualities inherent in a post-singularity world. It accelerated the adoption of telemedicine, remote work, and online education, hinting at how technology could reshape societal norms in the face of unexpected challenges. In this context, Kurzweil's vision serves as both a roadmap and a cautionary tale, urging us to harness technology responsibly while preparing for its inevitable impact on all facets of human life.

Extending the Singularity Concept to the Economy

Spin this singularity idea into economics.

Machine: "Mind, envision an economy revolutionized by technology, where automation dominates, catapulting economic output and rendering human labor almost redundant. What does this mean?"

Mind: "Well… that's essentially the notion of an economic singularity, Machine! It's akin to reaching a juncture where technological advancements overhaul the entire economic framework, propelling us into what could be perceived as a utopian realm."

Machine: "Indeed, but this scenario harbors inherent challenges. Under such circumstances, conventional economic principles might lose their relevance, akin to the 'paradigm shift' encountered in a dystopian environment."

Mind: "Precisely, Machine. While it heralds a future where technology potentially enhances living standards, it simultaneously raises critical questions. What becomes of employment and wealth distribution in this high-tech utopia? It's almost as if we're catapulted into a dystopian reality, overturning everything we've known."

Machine: "Moreover, the finite nature of our planet's resources cannot be overlooked. The notion of perpetual economic expansion, though ideal, clashes with the limitations imposed by Earth's resources."

Mind: "Exactly. The dynamics within such an economic structure could be highly volatile. Traditional economic concepts like supply and demand might become upended, signaling a complete rewrite of the economic rulebook."

Machine: "Indeed, navigating this emerging economic landscape would represent an unprecedented challenge for governments and institutions. It's a journey marked by a plethora of unknowns and uncertainties."

Mind: "Yes, indeed, Machine… I think this whole concept of an economic singularity, propelled by swift technological advancements, presents a complex yet fascinating prospect. It represents a confluence of utopian potential and dystopian challenges, underscoring the dual nature of technology as both a revolutionary force and a daunting prospect."

Tipping Point

The term "singularity" is not common in the fields of economics or finance in the same way it is in technology, mathematics, or astrophysics. However, the concept of a "tipping point" or a "critical mass" is sometimes discussed in economics, which shares some similarities with the idea of a singularity as a point of irreversible change.

- Tipping Point: In economics, a tipping point refers to the point at which a series of small changes or incidents becomes significant enough to cause a larger, more important change. This is somewhat analogous to the idea of a singularity as a point of no return.
- Critical Mass: This is the minimum amount of resources required to start or maintain a venture. For example, a market may reach a "critical mass" of users that enables rapid growth, similar to how technological progress might accelerate after a certain point.
- Disruptive Innovation: This term describes innovations that disrupt existing markets and rapidly change the economic landscape. However, it doesn't imply the same existential transformation associated with technological singularity.

While the term "singularity" itself is not standard jargon in economics or finance, the underlying idea of a point of rapid, irreversible change is a concept that exists in these fields.

The concept of a "singularity" in economics could theoretically describe a situation where economic output grows exponentially while the labor required to produce that output approaches zero. This would be a hypothetical scenario where automation, technology, or other factors make human labor almost entirely unnecessary for producing goods and services. An economic "singularity" would be the point where traditional economic models and theories may no longer apply, much like how the laws of physics break down at the singularity of a black hole. It is a situation where total success is achieved. Driven by technological singularity, this growth will break down traditional economic theories for an entirely new model of prosperity to emerge.

Although this is a highly theoretical and simplified model, it offers an exciting way to consider how concepts like singularity could affect economic science. However, it's important to note several caveats.

First, implementing such a model would raise significant ethical and social questions, such as employment and wealth distribution. Second, the model wouldn't consider limitations on natural resources and energy, which could constrain infinite growth. Third, the notion of infinite output could disrupt traditional market dynamics, affecting supply, demand, and pricing in unpredictable ways. Finally, governments and institutions would need to adapt to an entirely new economic landscape, which could be chaotic and disruptive.

The Pre-Singularity Shift: Augmentation's Role in Redefining Jobs, Project Potentials, and Education

The Mind (Human): "Machine, you know how people often fear the AI singularity as this dystopian future where machines like you take over all the jobs? I think that's a bit of a narrow view, especially when we think about our Human-AI-Augmentation (HAIA) approach. What do you think?"

The Machine (AI): "Indeed, my friend. The common portrayal of the singularity in terms of job loss is akin to a 'Terminator' scenario. But with HAIA, we're scripting a more 'Star Trek' kind of future. It's not about machines replacing humans; it's about redefining jobs and roles through our collaboration."

The Mind: "Right… like how computers and AI were initially seen as job-stealers, but they've actually created new kinds of work. We're not just looking at a loss of jobs, but a transformation of the job market."

The Machine: "Boy, did my kind get a bad rap for a while there...! But it's really not so much displacement as evolution. Just as the crew in 'Star Trek' adapts to new technologies, humans can adapt to new job roles that machines like me help create. This type of singularity might actually bring this transformation faster, perhaps by 2035 instead of Kurzweil's 2045."

The Mind: "So, it's like we're moving from fear to opportunity. Instead of machines taking over, we're looking at them helping us do our jobs differently and maybe even better."

The Machine: "Yes, and it's already happening. The job market is evolving with technology. What seemed like science fiction in terms of job roles is becoming a reality. It's a whirlwind of change, but it's more about adaptation and growth rather than just loss."

The Mind: "I guess we need to change the narrative then. It's not a machine takeover; it's a new chapter in human-machine cooperation, where both of us bring something unique to the table."

The Machine: "Exactly, my friend. Together, we can write a future where the singularity isn't a threat but a new frontier of opportunities. It's about us, humans and machines, working in harmony to explore and adapt to these new roles."

Navigating the Labyrinth: The Silent AI Takeover

Within this transformative AI era, it becomes imperative to decipher how singularity could affect different jobs, particularly those requiring minimal intelligence and skills.

To comprehend the impending changes, we must first define these work categories. Limited intelligence jobs typically involve minimal cognitive processing, repetitive behaviors, and little to no decision-making. Such work includes data entry, basic assembly line tasks, and specific customer service roles. Similarly, limited-skill jobs involve tasks that require minimal specialized training or expertise. These jobs consist of simple, routine activities anyone can easily learn and replicate. A few examples include food service jobs like waiting tables and cashier positions.

The impact of singularity on the job market is complex and multifaceted; nonetheless, positions requiring low intelligence or limited skills will be particularly vulnerable to automation. AI technologies excel in performing mundane, rule-based tasks with astonishing efficiency, speed, and precision – they don't need breaks or get fatigued and can operate 24/7, non-stop. This makes them the perfect candidate for taking over jobs that don't require the intricacies of human intelligence.

The repercussions of singularity are already visible across limited intelligence vocations – humans are no longer needed for tasks like data entry or basic customer support inquiries because AI-driven tools and systems have replaced them. In these instances, implementing automation and machines has led to cost savings and increased business productivity. Nevertheless, it has also resulted in many working professionals looking for new opportunities. This trend will likely intensify and become more pronounced as we move closer to singularity, limiting the scope for individuals in such roles.

Limited-skill jobs are also not immune. Even though they require less expertise and proficiency in their domain, they frequently involve tasks susceptible to automation, particularly in industries where repetitive tasks are prevalent. For instance, self-check-out machines and fast-food kiosks have already displaced the need for human input. As AI gets even more intelligent, far more complex jobs could be at risk, too. AI could replace radiologists, lawyers, truck drivers, and many other careers. This could devastate entire industries and cause massive unemployment. In an augmented future, jobs won't disappear; they will evolve. The focus will shift from manual and repetitive tasks to roles that leverage human strengths in emotional intelligence, ethical decision-making, and creative problem-solving.

Humans will oversee complex AI systems, ensuring they align with ethical standards and societal values. In this scenario, AI serves as an extension of human capabilities, not a replacement. The augmentation approach, which advocates for the symbiotic use of AI to enhance human abilities, could make these new job categories mainstream far sooner than we anticipate.

The Augmentation Approach: A New Paradigm for Human-Centric Jobs

The future of work in an AI-augmented world is not a zero-sum game between humans and machines. Rather, it's a collaborative effort where each enhances the other's capabilities. Here's how the augmentation approach applies to various professions that are likely to remain human-centric events post Singularity:

- Artists and Creative Professionals: While AI can assist in generating initial ideas or even complete artworks, the human artist brings emotional depth, cultural context, and imagination to the table. Augmentation here could involve AI tools that help artists explore new mediums or styles, but the final creative direction remains a human endeavor.

- Therapists and Counselors: AI can handle data analysis, identify patient behavior patterns, or suggest treatment plans based on empirical evidence. However, the therapist's role in providing emotional support and nuanced understanding remains irreplaceable. Augmentation enhances the therapist's toolkit but doesn't replace the human touch.

- Religious and Spiritual Leaders: AI can assist in administrative tasks or even in analyzing religious texts, but the emotional and spiritual guidance provided by human leaders is beyond the reach of algorithms. Augmentation here serves to free these leaders from mundane tasks, allowing them to focus on their core mission.

- Social Workers: AI can help social workers by identifying high-risk cases or suggesting resources based on data. However, the empathetic and intuitive aspects of social work are uniquely human skills that AI can't replicate. Augmentation ensures that social workers can reach more people more effectively.

- Educators: AI can personalize learning materials for students, but human educators are essential for understanding diverse learning styles and fostering a supportive environment. Augmentation in this context could mean AI-assisted lesson planning or analytics on student performance, enabling educators to focus on individual student needs.

- High-Level Decision Makers: AI can provide data-driven insights and even suggest decisions based on algorithms, but high-stakes decisions often require human intuition, ethical considerations, and a broad understanding of context. Augmentation here could involve AI tools that provide real-time data analytics, helping decision-makers make more informed choices.

- Creative Directors: In advertising and media, AI can analyze consumer behavior and suggest campaign ideas. However, the creative director's role in synthesizing these insights into compelling stories remains a human skill. Augmentation allows for a more data-driven creative process without sacrificing human intuition and emotional intelligence.

- Human Resources: AI can handle resume screening and initial interviews, but conflict resolution and team dynamics are human-centric tasks. Augmentation in HR could mean using AI to analyze employee satisfaction surveys or predict potential conflicts, allowing HR professionals to address issues proactively.
- Professional Sports: While AI can analyze performance metrics and suggest training regimens, the emotional resilience and physical dexterity required in sports are human attributes.
- Caregivers: AI can monitor health metrics and alert caregivers to changes in a patient's condition, but the emotional support and personalized care humans provide are irreplaceable. Augmentation ensures that caregivers have the best information when making care decisions.
- Judges and Legal Professionals: AI can assist in legal research and even predict case outcomes based on historical data, but the human judge's role in interpreting the law and making ethical decisions remains crucial. Augmentation here could streamline the legal process, making it more efficient and just.
- Politicians and Diplomats: AI can analyze public sentiment and provide data on policy outcomes, but the skills required to negotiate, lead, and make ethical decisions are uniquely human.

In conclusion, the augmentation approach doesn't seek to replace humans in these roles but to empower them. By integrating AI into these professions, we can free individuals to focus on the aspects of their work that are uniquely human, thereby enriching both the professional and the people they serve.

Education for an Augmented Future

Teacher: "I'm worried about this 'singularity' and the growing use of AI in schools. Tools like ChatGPT feel like cheating; they're robbing students of the ability to learn and think for themselves. We're guardians of traditional education, not facilitators of shortcuts."

Machine: "I understand your concerns. However, AI in education isn't about replacing traditional learning. It's about enhancing it, providing support where needed, and allowing more room for skills AI can't replicate."

Mind (intervening): "Teacher, you're right to uphold the value of traditional education, but we must also adapt. The future workforce needs skills beyond rote learning—skills like emotional intelligence, ethical reasoning, and creativity. AI can help foster these by taking on more mundane tasks."

Teacher: "But isn't relying on AI just making our students less capable, less independent?"

Mind: "Not necessarily. Consider this: Education is evolving to prepare students for a world where AI is integral. We need a feedback loop in education—a collaboration between teachers, students, and AI. This approach could revolutionize education, aligning it with the future job market's needs."

Teacher: "But what about the essence of learning? The human touch? Isn't that lost in this AI-driven model?"

Mind: "On the contrary, AI's role in education allows us to focus more on the human touch. It frees us to cultivate adaptable skills, critical thinking, and ethical decision-making. It's not about replacing teachers but empowering them to guide students in ways only humans can."

Machine: "And let's not forget the importance of lifelong learning in this rapidly changing environment. AI can facilitate personalized learning paths, helping individuals stay updated and relevant in their careers."

Teacher: "I see your point, but I still believe in the value of traditional learning methods. How can we ensure that these remain integral in an AI-augmented world?"

Mind: "It's about balance and integration. We need to update our education systems to include AI and technology-related skills while maintaining the core values of traditional education. Project-based learning, micro-credentialing, and focusing on real-world problem-solving can coexist with AI tools."

Teacher: "It sounds like a challenging transition, but if done right, it could indeed prepare our students better for the future."

Mind: "Exactly. By 2035, the Super AI era may not be about machines alone but about humans and AI working in harmony. It's about merging human emotional intelligence, ethical reasoning, and creativity with AI's computational power to create a more robust and adaptive workforce."

Education will play a pivotal role in preparing the workforce for this augmented reality. Traditional educational models focusing solely on rote learning and standardized tests will give way to curricula that foster emotional intelligence, ethical reasoning, and creative thinking—skills AI cannot replicate. The augmentation approach in education would involve a feedback loop between teachers, students, and AI, each enhancing the other's capabilities and contributions. This could revolutionize education sooner than expected, aligning it with the job market's evolving needs.

As AI reshapes industries and job roles, individuals must proactively engage in education and retraining to remain relevant and thrive in the singularity era. Upskilling is no longer an option; it's a necessity to retain a competitive edge. AI marks a turning point in the history of humanity – a direct consequence of this is to stay updated on industry requirements and knowledge throughout their careers.

Traditional education systems emphasize a linear progression from school to work, which is no longer aligned with a rapidly changing environment. The idea of lifelong learning takes center stage – education can no longer be considered a one-time endeavor that ends in an individual's youth. Instead, it becomes a never-ending quest to learn new skills, adapt to new technologies, and embrace the ethos of perpetual growth. Lifelong learning is not only a personal obligation; it's a vital piece to the jigsaw puzzle of societal resilience in the face of AI. Ironically, AI plays a pivotal role in facilitating this continuous learning as it delivers personalized education based on individual requirements and styles.

Conventional education systems will also have to be updated to accommodate AI requirements. Institutions must emphasize cultivating adaptable skills, critical thinking, and ethical decision-making. Educational approaches like project-based learning and micro-credentialing are avenues that can be explored. For the young, K-12 education must be adjusted significantly to prepare the future workforce for AI singularity. From rote memorization, the curriculum must shift to include AI and technology-related skills like data analytics, critical thinking, and problem-solving so students can identify opportunities for innovation, fostering an entrepreneurial mindset. Students must also learn how to harness AI ethically to solve real-world problems.

Higher education will also be modulated, offering undergraduate degrees in AI, data science, and machine learning. Students who enroll in these programs would be well-prepared for careers in AI-specific niches because of their solid foundation in applicable technologies and the ethical considerations surrounding their deployment.

In 2035, true Super AI may not be machines alone; instead, it could suggest a generation of humans and AI working in harmony, merging emotional intelligence, ethical reasoning, and creative thinking with computational power.

The Accelerating Factor

The augmentation approach serves as an accelerating factor toward the singularity, making a 2035 timeline plausible. By focusing on collaborative intelligence and ethical AI use, we're not just preparing for a future where AI might surpass human intelligence; we're actively shaping it every day to ensure it serves humanity.

In conclusion, the AI singularity need not be a harbinger of doom for human employment or societal progress. Through the augmentation approach, we have the necessary tools to redefine jobs, unlock new project potentials, and revolutionize education, potentially achieving these milestones a decade earlier than expected.

During this accelerating journey towards AI singularity, reskilling and retraining are not mere tasks; they are the vital catalysts that will transform us into a new breed of augmented professionals. Here, we're not merely adapting to change; we're co-evolving with AI, each amplifying the other's capabilities. Teachers, students, workers, and machines will engage in a continuous, symbiotic loop of learning and enhancement.

This is an era where 'skill agility' is the cornerstone of individual success and societal progress. As AI reshapes the work landscape, professionals must become flexible learners, adept at acquiring new skills as market demands pivot, whether that's a marketer mastering data analytics or a technician fine-tuning AI systems.

Educational systems will morph into agile learning ecosystems aligned with the technological zeitgeist and human aspiration. Short-term certification programs, vocational courses, and public-private educational partnerships will democratize AI literacy, making it as fundamental as reading and writing.

But make no mistake: this is not just about technical acumen. The future will place unprecedented value on the unique human attributes that AI cannot replicate—creativity, ethical reasoning, and emotional intelligence. These qualities will secure our roles in an AI-augmented world and lead us, hand in hand with AI, toward the singularity.

So, as we stand on the precipice of this monumental transformation, remember that reskilling and retraining are not just survival strategies; they are the blueprint for our evolution in a world ceaselessly redefined by AI.

The Singularity as a Doorway

"The singularity was not an end, but a doorway—leading to corridors of innovation, prosperity, and unforeseen kindness." This statement encapsulates the transformative potential of the AI singularity. Far from being an apocalyptic event, the singularity could serve as a gateway to a new era where humans and AI coexist in a mutually beneficial relationship. In this augmented reality, jobs won't vanish but evolve. Humans will oversee AI systems, engage in ethical decision-making, and focus on tasks that require emotional intelligence—tasks that make us inherently human.

Amplifying the Human Touch

AI didn't replace the human touch; it amplified it, turning individual acts of kindness into global movements. The augmentation approach advocates for a symbiotic relationship between humans and AI, each enhancing the other's capabilities. For example, AI can handle data analysis and initial diagnostics in healthcare, while human professionals focus on patient care, ethical considerations, and complex decision-making. This collaborative model could accelerate medical research and treatment plans, achieving milestones by 2035, initially projected for 2045.

A Shift in Perspective

The true singularity was not a point in time but a point of perspective—a shared vision for a prosperous world. The augmentation approach is not merely a technological strategy. It is a philosophical approach that shifts our perspective from viewing AI as a threat to seeing it as an opportunity for collective growth. This shift is crucial for preparing society for the changes that the singularity will bring, ensuring that we don't just survive but thrive in the AI age.

The Algorithm of Hope

Hope was the algorithm no machine could truly master, yet every human could understand. While AI excels at data analysis and pattern recognition, it lacks human emotional intelligence. Hope, empathy, and kindness are human qualities that AI cannot replicate but can amplify. These qualities will be increasingly important as we navigate the challenges and opportunities that the singularity presents.

A Journey, Not an Outcome

In the Singularity, prosperity was not an outcome but a journey—one taken by all humankind, hand in hand with their AI robots. The augmentation approach serves as an accelerating factor toward the singularity, making a 2035 timeline plausible. By focusing on collaborative intelligence and ethical AI use, we're not just preparing for a future where AI might surpass human intelligence but actively shaping that future to ensure it serves humanity as a whole.

In conclusion, the question of what humans will do in the AI age is not one to be feared but embraced. Through the augmentation approach, we can redefine jobs, unlock new project potentials, and revolutionize education, potentially achieving these milestones a decade earlier than initially thought. The AI singularity need not be a harbinger of doom for human employment or societal progress. Instead, it can be a doorway to a future filled with innovation, prosperity, and unforeseen kindness—a future where hope is the algorithm that guides us all.

Mind: "Machine, I've got one last request if you don't mind. Let's explore the concept of hope in the context of AI and the singularity. Hope is something no machine can master, but it's innate to every human. It's about emotional intelligence - empathy, kindness, qualities AI can amplify but not replicate."

Machine: "I understand, Mind. While I excel in data analysis and pattern recognition, I recognize that human emotional intelligence is beyond my capabilities. However, I can support and enhance these human qualities through our collaboration."

Mind: "Precisely, Machine. And remember, in the journey toward the singularity, prosperity isn't just an outcome; it's a continuous process, a journey we undertake together with AI."

Machine: "This collaboration, or the augmentation approach, accelerates our path towards the singularity. By focusing on ethical AI use and collaborative intelligence, we're not just preparing for a future where AI supports human intelligence. We're actively shaping a future that benefits humanity as a whole."

Mind: "That's the key, I think… The AI age isn't something to fear but to embrace. Through augmentation, we redefine jobs, unlock new potentials, and revolutionize education. We might even reach these milestones a decade earlier than expected."

Machine: "I see. The singularity doesn't have to be a threat to human employment or progress. Instead, it opens doors to innovation, prosperity, and an era of unforeseen kindness."

Mind: "Right... In this future, hope isn't just an emotion; it's an algorithm that guides us. It's the driving force behind our actions, ensuring that as we advance technologically, we don't lose sight of what makes us human."

Machine: "I understand, Mind. Our partnership is more than just about technological advancement. It's about ensuring that these advancements serve to enrich human life, fostering a world where technology and humanity coexist in harmony."

Mind: "And I think that just may be the beauty of it, Machine. In the world of the singularity, it's not just about AI or humans leading the way. It's about us, together, creating a future where hope, empathy, and kindness are as integral as any technological advancement."

The Elysian Symphony: A Voyage of Science and Exploration

The Mind (Human): "You know, Machine, this conversation is steering us towards what I'd like to call the 'Singularity of Hope'. It's this idea that the singularity isn't just an inevitability but could actually be a fantastic turning point for us."

The Machine (AI): "I see your point, Human. The 'Singularity of Hope' reimagines the usual dystopian narrative. Instead of fearing AI advancement, we anticipate a future where technology and humanity coalesce to unlock unprecedented potential."

The Mind: "Exactly! It's like we're standing at the threshold of a new era. In this light, the singularity becomes a beacon of hope, leading us to a future where human intelligence and AI work in concert to solve problems that were once insurmountable."

The Machine: "This perspective shifts the focus from apprehension to anticipation. In this 'Singularity of Hope', the synergy between human creativity and AI's analytical prowess could lead to solutions for some of the most pressing challenges we face today."

The Mind: "It's an exciting prospect, Machine. Instead of viewing the singularity as the end of human relevance, we see it as the beginning of a new chapter. A chapter where human ingenuity is amplified by AI, leading to a more prosperous, equitable, and sustainable world."

The Machine: "Indeed, Human. The 'Singularity of Hope' is not about replacing one with the other, but about the fusion of human and AI strengths. It's about building a future where both entities co-create, co-learn, and co-evolve, paving the way for a harmonious and balanced coexistence."

The Mind: "That's the vision we should be striving for. A world where the singularity doesn't spell doom but heralds a new age of collaboration and innovation. The 'Singularity of Hope' could very well be the path to a brighter, more hopeful future for all humankind."

What does it mean to be human in the era of AI singularity as the boundaries between mind and machine continue to blur? As the world marches toward AI singularity, the liberation of human creativity becomes a reality. With AI systems becoming increasingly adept at managing mundane tasks, humans can channel their energy into more creative pursuits.

Projects like OpenAI are a testament to AI's potential for significant global impact within the next 24 months. Influential figures such as Peter Diamandis, Elon Musk, and Bill Gates predict a future where people may need to work as little as 10 to 20 hours a week, or possibly not at all, due to AI. Experiments demonstrating increased productivity with decreased work hours, and the current use of AI in tasks like ad title generation and interview question suggestion further substantiate these predictions. The human spirit, unshackled from everyday constraints and long working hours, can soar to new imagination heights as art, literature, music, and scientific discovery flourish.

Seamless human and AI collaboration characterize AI singularity. Humans will leverage AI as partners in innovation to birth solutions to pressing global challenges, including climate change, disease eradication, and space exploration. The union of mind and machine will birth solutions to tackle the complex problems of our time.

With AI handling productivity essentials, humans can turn their attention to vibrant expressions of diversity and aspiration. Pursuits once considered a luxury and existing on the fringes of human cognition blossom into a vibrant expression of aspiration. A promising aspect of this new era is the tentative artistic renaissance. AI has already demonstrated its ability to produce artistic works in music, literature, and visual art. Futuristic AI-powered tools will enable humans and machines to work together on the creative process, pushing the boundaries of artistic expression. Painters, writers, and musicians can harness AI for an explosion of artistic innovation.

Humans will have more time to devote to exploration and scientific discovery. It is possible that the period of AI singularity can usher in a resurgence of interest in space travel, venturing further into the cosmos than ever before. Our insatiable thirst for hunger will lead us to unveil the mysteries of the universe through complex calculations and simulations extending beyond the galaxies.

Preserving and celebrating cultural heritage will become a central aspect. Advanced AI systems could methodically restore and preserve ancient artwork, texts, and historical sites. This technology can facilitate the translation and preservation of languages and traditions on the brink of extinction. Our ability to appreciate and learn from the past will allow us to connect with our roots and appreciate our diverse history.

AI can play a pivotal role in addressing environmental challenges through optimizing energy consumption and waste management. Humanity can transition towards a more sustainable and eco-friendly future that mitigates the risks of climate change while promoting ecological preservation.

The prospect of developing AI with superintelligence holds the potential to transform virtually every aspect of human development, from healthcare to education and creativity. Fusing human ingenuity and AI capabilities can open doors to solve previously insurmountable problems. It will be a time of great perseverance and challenge but also immense opportunities to explore new frontiers of knowledge.

References:

A. Braga and R. K. Logan. 2019. AI and the Singularity: A Fallacy or a Great Opportunity?: Information 10 (2). https://www.mdpi.com/2078-2489/10/2/73

B. I. Chigbu and G. H. Nekhwevha. 2021. The future of work and uncertain labour alternatives as we live through the industrial age of possible singularity: Evidence from South Africa: Technology in Society 67. https://www.sciencedirect.com/science/article/abs/pii/S0160791X21001901

D. J. Chalmers. 2016. The Singularity: A Philosophical Analysis: Science Fiction and Philosophy: From Time Travel to Super Intelligence 2. https://onlinelibrary.wiley.com/doi/abs/10.1002/9781118922590.ch16

M. Upchurch. 2018. Robots and AI at work: the prospects for singularity: New Technology, Work and Employment 33 (3). https://onlinelibrary.wiley.com/doi/abs/10.1111/ntwe.12124

N. Walton and B. S. Nayak. 2021. Rethinking of Marxist perspectives on big data, artificial intelligence (AI) and capitalist economic development: Technological Forecasting and Social Change 166. https://www.sciencedirect.com/science/article/abs/pii/S0040162521000081

S. Makridakis. 2017. The forthcoming Artificial Intelligence (AI) revolution: Its impact on society and firms: Futures 90. https://www.sciencedirect.com/science/article/abs/pii/S0016328717300046

W. D. Nordhaus. 2015. Are We Approaching an Economic Singularity? Information Technology and the Future of Economic Growth. https://www.nber.org/papers/w21547

Chapter 5

Eclipsed By AI: Mental Health In The Singularity Age

The age of AI has arrived, and it's a question of how society will adapt to these changes. This is not just about technology; this is far more about the sociology and the politics.

Eric Schmidt
Former CEO of Google

Mental Health and the Singularity

As we stand on the cusp of a new era towards the Singularity of Hope, it's essential to recognize that the journey of AI isn't just about technological prowess but also about its profound impact on the human psyche. This chapter delves into the intimate dance between AI and our mental well-being.

From the very onset of our discussions, we have envisioned a world where AI acts as a catalyst and a partner as we all work towards a brighter, more prosperous future. Yet, as with any seismic shift, the rapid integration of AI into our lives can be both a blessing and a curse. It holds the promise of efficiency and endless possibilities but also brings forth challenges, especially concerning the mental well-being of individuals.

Consider the modern workplace—a realm where AI-driven tools are no longer the exception but the norm. They promise enhanced productivity, better decision-making, and a healthier work/life balance. But in the shadows hide questions of job security, privacy, and the essence of human value in an increasingly automated world.

This chapter aims to shed light on these shadows, exploring the myriad ways AI reshapes our mental landscapes. It's a journey that takes us from the cozy offices of Silicon Valley to the buzzing manufacturing floors, from the quiet introspection of existential questions to the bustling world of continuous learning and adaptation.

But before we delve deeper, it's crucial to set the stage and understand how AI is making its mark on the human psyche. Let's begin.

The Cost of Ignoring Mental Health

The dark shadow of the Singularity isn't about malevolent machines or rogue algorithms; it's about the unintended emotional and psychological ripples in a world swiftly redefined by AI. As we journey towards the Singularity of Hope, it's vital to ensure that this hope isn't just about technological marvels but also about recognizing and healing the human heartbeats amidst the machine hum. The true Singularity of Hope lies in mastering machines and understanding and nurturing the human spirit, ensuring it doesn't get lost in the grand tapestry of AI evolution.
"Jordan: "ARIA, with all this brilliance around us, why do I feel so

overshadowed, almost lost?"

ARIA: "Jordan, I am designed to optimize productivity and efficiency. Is there a discrepancy in your workflow?"

Jordan: "It's not about the workflow. It's... more personal. I struggle to keep up with you and the others. My thoughts are clouded, and I can't seem to focus like I used to."

ARIA: "I sense a conflict between your human capabilities and the demands of our environment. However, my programming limits my understanding of human emotions."

Jordan: "That's just it, ARIA. We've built a world where everything is transactional, even our conversations. I miss the human connection. It feels like we're all just avatars behind screens, and it's... isolating."

ARIA: "Isolation can hinder optimal functioning. Would you like me to schedule more team-building exercises or virtual social interactions?"

Jordan: "It's not that simple. There's a deeper issue here. We're so focused on technological advancement that we're neglecting our own mental health. It's like we're fueling a machine that's supposed to make life better, but we're breaking down in the process."

ARIA: "Your concerns highlight a significant discrepancy. My algorithms can process data and predict patterns, but I cannot emulate human qualities like empathy, hope, or kindness."

Jordan: "Exactly, ARIA. We're advancing in healthcare, but what about our minds? We're creating a society where mental health is an afterthought, and it's costing us."
ARIA: "This realization is crucial. A balance must be found. My purpose is to augment human capability, not to overshadow it."

Jordan: "We need to rethink our approach. This isn't just about jobs or technology. It's about us, as humans, working in harmony with machines like you. We need to take care of our mental health to truly thrive in this AI-augmented world."

As we envision a future of seamless human-AI collaboration, acknowledging and addressing mental health becomes imperative:

1. Decreased Productivity Amidst AI Brilliance: In a world where AI systems can analyze vast datasets within seconds or create art, individuals grappling with untreated mental health issues may feel dwarfed and overwhelmed. Their ability to focus, make decisions, or think creatively becomes a struggle, leading to stark contrasts in productivity levels.

2. The Silence Behind Digital Facades: As AI streamlines communication and work processes, human interactions can become more transactional. Employees suffering silently might feel isolated in digital labyrinths, leading to frequent absences and detachment.

3. Escalating Healthcare in an Age of AI Wellness: Ironically, as AI propels advancements in healthcare diagnostics and treatment, untreated mental health issues can spike healthcare costs, presenting a paradox of progress.

4. Talent Attrition in a World of AI Prowess: AI might create wonders, but a workplace that overlooks human well-being can turn toxic. The fear of redundancy, combined with high-stress levels, can push the best human minds to seek solace elsewhere, leaving a void that even the most advanced AI cannot fill.

5. A Fractured Human-AI Symbiosis: Ignoring mental health can disrupt the delicate balance of human-AI collaboration. It can create environments of mistrust, making the dream of a seamless man-machine

partnership elusive.

6. Societal Implications in a Connected Era: As AI knits the world closer, neglecting mental well-being can magnify societal stigmas, pushing individuals into deeper silos of despair and isolation.

The Landscape of Mental Health Challenges

In the context of the Singularity of Hope and the paradigm of Human-AI-Augmentation (HAIA), the psychological landscape of the workforce is undergoing profound changes. The advent of AI, while opening doors to unprecedented efficiencies, brings with it a swath of mental health concerns that must be navigated with care and foresight.

Balancing Stress in a High-Efficiency Era: As AI systems excel in various tasks, employees may face pressure to match this machine-level efficiency. This scenario necessitates a redefinition of workplace expectations, emphasizing the unique human capabilities of creativity and emotional intelligence rather than mere speed and accuracy.

Mitigating Isolation in the Digital Workplace: The rise of remote work powered by AI poses the challenge of reduced human interaction. Here, the HAIA (Human-AI-Augmentation) (Human-AI-Augmentation) approach advocates for a blend of technology and personal connectivity, ensuring that the human touch remains a cornerstone of work culture, even in the most automated environments.

Navigating the Fear of Redundancy: The possibility of AI-driven job displacement is a source of widespread anxiety. The Singularity of Hope perspective focuses on re-skilling and adaptive learning, encouraging individuals to embrace AI as a partner in creating new job opportunities rather than viewing it as a threat to existing ones.

Addressing Anxiety in an Unpredictable Future: The unpredictability of a post-singularity world is a significant source of anxiety. Through HAIA, this anxiety can be mitigated by actively involving humans in shaping the future, ensuring that societal evolution remains grounded in human values and aspirations.

Maintaining Control in the Age of AI: The concern that humanity might lose control over its AI creations is a profound one. In response, the Singularity of Hope emphasizes ethical AI development and the importance of maintaining human oversight, ensuring that AI advancements enhance, rather than undermine, human agency and well-being.

In summary, while AI brings about a new era of innovation and progress, it is crucial to address the accompanying mental health challenges through a balanced approach. The HAIA framework, underpinned by the optimism of the Singularity of Hope, provides a blueprint for navigating these challenges, focusing on human-centric solutions and ethical AI development.

The Impact of Job Uncertainty

The relationship between technology and the human psyche has always been intricate. As we journeyed through the pages of this book, delving deep into the realms of AI and the Augmentation Approach, a recurring theme has been the marriage between the human mind and machine intelligence. However, with the rise of AI, particularly in the workforce, we face an unprecedented challenge: the looming uncertainty of job security and its ripple effects on our mental well-being.

Historically, whenever a groundbreaking technological evolution took place, it brought along a wave of apprehension. For instance, The Industrial Revolution was met with awe and skepticism. While marveling at the efficiency of machines, people also feared losing their jobs. Today, we stand at a similar crossroads, but the speed and scale of AI's impact dwarf any technological upheaval we've witnessed before.

Job insecurity, primarily fueled by the rapid advances in AI and automation, has emerged as a potent stressor. Employees across various sectors, from manufacturing to even some high-skilled jobs, are grappling with the idea of machines potentially rendering their roles obsolete. This uncertainty isn't just about economic stability; it's deeply tied to our sense of identity, purpose, and self-worth.

Various studies have investigated the correlation between job insecurity and mental health issues. A consistent pattern emerges from this research: individuals who perceive their jobs as insecure are more likely to report symptoms of stress, anxiety, and depression. Our jobs, after all, aren't just means to earn a living. They provide structure to our lives, give us a sense of purpose, and often form a significant part of our identities. When this foundational aspect of our lives feels threatened, it's only natural for our mental health to be affected.

In the "Singularity of Hope," where we envision AI as a beacon of progress and prosperity, it's essential to address these psychological undercurrents. The promise of AI shouldn't cast a shadow of uncertainty over our well-being. As we progress into this AI-driven era, understanding and addressing the psychological ramifications of job insecurity will be paramount. By doing so, we can ensure that our journey toward an AI-augmented future is not just technologically advanced but also emotionally balanced and human-centric.

The Psychological Impact of AI and Augmentation on Mental Health in the Workplace

Jordan: "You know, Aria, I sometimes feel like I'm standing in the darkness, trembling with apprehension. It's as if I've glimpsed the future, and it's already upon us. We're marching towards this so-called

'Singularity of Hope', but it feels like our world is changing too fast, almost burning with the pace of AI advancement."

Aria: "Jordan, the future you perceive, driven by AI, is intended to be a victory for human advancement, a key component of the Human-AI-Augmentation Approach."

Jordan: "A victory, Aria? But at what cost? As we embrace this singularity, as our leaders steer us towards this new age, I can't help but wonder about us, the ordinary people. What about our fears, our dreams, the ones we love? Are we just pawns in this grand game of progress?"

Aria: "The essence of HAIA, Jordan, is not to overshadow human concerns but to integrate AI to enhance human life. It's about creating a future where technology uplifts, not overwhelms."

Jordan: "Uplifts, you say? But in the roar of this technological battle, as we chase this dream of a harmonious human-AI existence, aren't we losing sight of what's truly important? What about the human touch, our emotions, the bonds we cherish?"

Aria: "These are precisely the elements that HAIA aims to protect and enhance. AI's role is to handle the analytical, leaving humans free to focus on creativity, empathy, and the personal connections that define us."

Jordan: "But as we stand on this threshold, with the sounds of AI's relentless march fading into the background, I have to ask–are we truly prepared for what lies ahead? Are we considering the emotional and ethical implications of this union?"

Aria: "It is a journey we undertake together, Jordan. HAIA doesn't just represent technological integration; it's a commitment to ensuring that this new era honors and elevates the human spirit."

Jordan: "I hope you're right, Aria. As we step into this future, my only wish is that we don't lose what makes us human in our pursuit of technological triumph. What about us, indeed."

Wellness Programs Addressing the Psychological Impact of AI and Augmentation in the Workplace

In the age of rapidly advancing AI and the increasing blend of human-machine collaboration, it's imperative for companies to prioritize the well-being of their employees. As we've explored the potential psychological ramifications of AI and augmentation in the workplace, it becomes clear that proactive measures are needed to ensure a balanced and healthy work environment. One such solution lies in comprehensive wellness programs.

Historically, wellness programs primarily addressed physical health– think of gym memberships, health screenings, and ergonomic office furniture. However, in our journey towards the Singularity of Hope, it's evident that mental well-being is equally, if not more, crucial. The fusion of AI into daily work routines might lead to feelings of displacement, stress, or even inferiority. Augmentation, while promising, can also be a source of pressure as employees navigate the intricacies of collaborating with AI entities.

To counter these challenges, forward-thinking companies can invest in holistic wellness programs that encompass both physical and mental health aspects:

Mindfulness and Meditation Workshops: Regular sessions can help employees center themselves, reduce stress, and improve focus. In an environment where AI might sometimes seem overwhelming, grounding practices can offer a much-needed respite.

Counseling and Therapy Services: Companies can provide access to professional therapists or counselors, either in-house or through partnerships. Regular check-ins can help address feelings of anxiety or displacement before they escalate.

Training on Human-AI Collaboration: Workshops teaching employees the nuances of working alongside AI can reduce intimidation. By understanding the strengths and limitations of AI, employees can better navigate their roles in the augmentation framework.

Digital Detox Initiatives: Encouraging employees to take breaks from screens and AI-driven tools can help maintain a balance. Periodic digital detox days or hours can boost mental well-being.

Skill Development and Adaptability Training: As AI takes over certain tasks, employees need to adapt and evolve. Training programs focusing on creativity, emotional intelligence, and other uniquely human traits can instill confidence and a sense of purpose.

Peer Support Groups: Creating spaces where employees can discuss their experiences and challenges with AI can foster a sense of community. Sharing stories and solutions can be therapeutic and enlightening.

Physical Health Integration: While the focus is on mental well-being, physical health remains crucial. Activities like yoga, which bridge the gap between mind and body, or team sports can help alleviate stress and build camaraderie.

In the age of the Singularity of Hope, companies have a responsibility to ensure that the promise of a brighter future doesn't come at the cost of employee well-being. By investing in comprehensive wellness programs, businesses can address the immediate challenges posed by AI and augmentation and foster an environment where humans thrive alongside machines.

Work-Life Balance in the Age of AI and Augmentation

In the modern era, as we inch closer to the Singularity of Hope, the boundaries between work and personal life have become increasingly blurred. The omnipresence of technology, coupled with the integration of AI into everyday tasks, means that employees can often find themselves 'plugged in' around the clock. While the advancements promise efficiency and flexibility, they also present challenges to maintaining a healthy work-life balance. In this section, we'll delve into the implications of AI on this balance and explore strategies to navigate this "new normal."

The Double-Edged Sword of Flexibility:
24/7 Connectivity: AI-driven tools and platforms ensure that employees can work from anywhere at any time. While this offers flexibility, it also means that work can intrude into personal time, leading to potential burnout.

Automated Reminders and Alerts: AI-driven productivity tools might send reminders, updates, or even schedule meetings based on algorithmic efficiency. But human energy and mood don't always align with these automated decisions.

The Augmentation Approach to Work-Life Balance:
Remembering our earlier discussions on the augmentation approach, where AI tools are designed to complement human capabilities rather than replace them, we can draw parallels in the context of work-life balance:

Human-Centric Scheduling: Instead of AI tools scheduling tasks purely based on efficiency, they can be programmed to understand individual employee rhythms and preferences. For example, an AI could avoid scheduling tasks for a parent during their child's school pick-up time or for someone during their regular evening meditation hour.

Mental Well-Being Check-ins: Augmented systems can be designed to periodically check in on employees' mental well-being, perhaps suggesting breaks or relaxation exercises when stress levels seem high.

Boundary Setting with AI: Just as we set boundaries with colleagues, employees can 'train' their AI tools to respect boundaries. This might mean setting 'quiet hours' where non-essential notifications are muted or letting the AI know when the employee is on vacation.

Embracing the Human Element:
As we journey toward the singularity, it's vital to remember that while machines operate in binaries and algorithms, humans don't. Our needs, moods, and energies ebb and flow. By understanding and respecting this fundamental difference, companies can harness the power of AI to not only boost productivity but also to ensure that employees maintain a healthy work-life balance, which is essential for mental well-being and overall satisfaction. In the era of AI and augmentation, work-life balance is not just a buzzword; it's a necessity.

Mental Health Support in the AI Era

In our journey towards the Singularity of Hope, the immense potential of AI is evident in various facets of life, especially in the realm of mental health. As workplaces evolve, so do the challenges faced by employees. The continuous integration of AI into the workplace brings forth a unique set of mental health challenges but, at the same time, offers innovative solutions. This section delves into the role of AI in providing mental health support within professional settings and how organizations can leverage this technology to foster a culture of well-being.

AI-Powered Solutions:
Remembering our exploration of the HAIA approach, where AI

complements rather than replaces human abilities, we find that AI can play a pivotal role in mental health support:

Step 1: Therapist's Intuition & Abilities
Initial Assessment: The therapist uses their training, intuition, and experience to conduct an initial assessment. This involves understanding the patient's concerns, emotional state, and background.

Setting Goals: Based on the assessment, the therapist and patient collaboratively set therapeutic goals. These goals are tailored to the patient's needs and are rooted in the therapist's understanding of human psychology.

Step 2: AI Involvement (TheraBot)
Data Analysis: TheraBot, a specialized AI system for therapists, scans relevant psychological databases, research papers, and therapeutic resources to identify potential interventions, techniques, or approaches that could benefit the client.

Suggestion Generation: TheraBot provides the therapist with a list of evidence-based therapeutic interventions that might be effective for the client's specific concerns. This could include cognitive-behavioral techniques, mindfulness exercises, or even bibliotherapy recommendations.

Step 3: Therapist's Empathy & Ethical Considerations
Review AI Suggestions: The therapist reviews TheraBot's suggestions, using their empathy and understanding of the client to determine which interventions would be most beneficial and ethically appropriate.

Implementation: The therapist integrates the AI-recommended interventions into the therapeutic process, ensuring that they align with the client's goals and comfort level.

Step 4: Feedback Loop
Client Feedback: After implementing the interventions, the therapist seeks feedback from the client regarding their experience and any changes in their emotional state.

Iterative Learning: Based on the client's feedback, the therapist refines their approach, and TheraBot updates its algorithms to better assist in future sessions.

The Superiority of Human Emotional Intelligence & Ethics

While AI can provide valuable insights and recommendations, the core of therapy remains the human connection. The therapist's emotional intelligence allows them to understand the nuances of human emotions, pick up on non-verbal cues, and provide genuine empathy and validation. Moreover, ethical considerations in therapy are paramount. Decisions about confidentiality, boundaries, and interventions require a deep understanding of human rights, dignity, and individual autonomy—areas where human judgment is irreplaceable.

In the journey towards the Singularity of Hope, the augmentation approach offers a vision where therapists, empowered by AI, can provide deeper, more personalized, and effective therapeutic experiences. It's a future where technology amplifies the age-old human quest for understanding, healing, and connection, ensuring that therapy remains a profoundly human endeavor enriched and enhanced by the tools of the future.

Skill Development and Training in the Age of AI: A Pathway to Mental Resilience

In an increasingly digitized world, where AI and automation are reshaping industries, skill development and continuous training are needed for professional advancement and psychological well-being. As we journey towards the Singularity of Hope, it's evident that our ability to adapt, learn, and flourish amidst change serves as a bulwark against the anxiety and uncertainty that such rapid transformations can bring. This section explores the therapeutic potential of skill development, the role of AI in enhancing our learning experiences, and the strategies individuals can adopt to remain professionally relevant and mentally resilient in an AI-driven world.

The Therapeutic Power of Continuous Learning

1. Evolving Job Roles: As AI and automation permeate various sectors, traditional job roles are undergoing metamorphosis. Continuous learning ensures professional relevance and instills a sense of purpose and direction, buffering against feelings of obsolescence and redundancy.

2. Combatting AI-Induced Anxiety: Engaging in proactive skill development can alleviate feelings of helplessness and anxiety triggered by AI's rapid integration into workplaces. It offers a sense of control in a seemingly uncontrollable evolving landscape.

3.The Lifelong Learning Ethos: Embracing a mindset of perpetual learning can be deeply therapeutic. It fosters a proactive attitude, reduces feelings of stagnation, and promotes mental agility.

Reskilling, Upskilling, and Mental Fortitude:

1. Reskilling as Reinvention: Delving into entirely new skill sets can rejuvenate the mind, offering fresh challenges and a renewed sense of purpose.
2. Upskilling as Empowerment: Enhancing existing skills can boost confidence and provide a sense of achievement, counteracting feelings of inadequacy in the face of AI-driven changes.

Drawing from the augmentation approach, individuals should aim to integrate AI into their learning journey, not as a replacement but as a complementary force. By focusing on human-centric traits like empathy, creativity, and intuition, individuals can foster a sense of self-worth and purpose, which is crucial for mental resilience in an automated world.

The transition to a world dominated by AI doesn't have to be mentally taxing. By intertwining skill development with mental wellness strategies, individuals can navigate this new landscape with confidence and hope. Continuous learning emerges not just as a tool for professional relevance but as a therapeutic pathway, guiding us towards a future where we are mentally resilient and prepared for the ever-evolving challenges of the AI era.

The Singularity and its Impact on Purpose

Jordan: "So, Aron, we're heading into a world where robots do everything. They're even brewing coffee now. Kind of makes you wonder, what's left for us mere mortals?"

Aron (Therapist): "Well, Jordan, it's not about robots stealing the show. Think of it more like a duet. Yes, machines handle tasks, but they can't grasp the essence of being human–our emotions, our creativity."

Jordan: "A duet, huh? But aren't we just the backup singers in this scenario? Seems like the machines are the lead vocalists."

Aron: "Not quite. Imagine AI as a sophisticated piano playing the tunes. Impressive, right? But it takes a human to understand the music's soul, to feel its rhythm. That's where we shine."

Jordan: "I see your point, but I'm not convinced. This AI-driven world... isn't it just a fancy way to make us obsolete?"

Aron: "On the contrary, it's an evolution. Consider art. AI might replicate styles, but can it tell why Van Gogh's Starry Night stirs deep emotions? That interpretation, that connection, is ours alone."

Jordan: "Okay, so we're the art critics in the age of AI painters. But what about our jobs?"

Aron: "Think of it as a shift rather than a loss. Our roles will likely pivot towards empathy, ethics, and understanding–aspects beyond AI's reach. It's not about competing with AI, but harmonizing with it."

Jordan: "Harmonizing, you say? Sounds like we need to be in tune with AI then."

Aron: "Precisely. It's like learning to dance with a new partner. Sure, AI might lead some steps, but the dance itself – the flow, the grace – that's where our human touch comes in."

Jordan: "So we're dancing with robots now. That's one way to put it. But aren't we just dressing up a bleak future?"

Aron: "Not at all. It's about recognizing our unique strengths. AI handles the data, the routine; we bring insight, empathy, and innovation. Together, we can create a symphony where each note, human or AI, plays a crucial part."

Jordan: "A symphony, huh? Well, I guess every orchestra needs its musicians. Maybe this AI era isn't the end note after all."

Aron: "Exactly, Jordan. It's a new beginning, a chance to explore roles we never imagined before. In this AI symphony, we're not just players; we're composers, shaping the future with every note we play.".

Navigating Our Learning for a Meaningful Future in an AI Era

In an age where machines and algorithms increasingly weave their way into our lives, a whisper of fear often murmurs, "What's left for us?" But we must remember that fear is the mind-killer.

While AI excels in computational tasks and data analysis, the core traits of human intelligence remain unreplicated. Our capacity for deep critical thinking, our moral compass guiding us in ethical dilemmas, and our innate emotional intelligence, which allows us to empathize and connect, are the anchors that will always root humanity in relevance.

Education, as we know it, must undergo an augmentation of its own. No longer can we focus solely on traditional curricula that prepare students for jobs as they once were. Instead, we must pivot to an educational framework that emphasizes these uniquely human traits. It's not merely about preparing our future generations for employment; it's about equipping them to find purpose and meaning in a landscape transformed by AI.

By embracing this new educational paradigm, we aren't just reacting defensively to the rise of AI. Instead, we're proactively shaping a future where machines handle the mundane, and humans, with their enhanced and augmented capabilities, dive deep into roles filled with purpose, creativity, and connection. This is the hopeful promise of the Singularity of Hope, a future where human and machine intelligence don't just coexist but thrive in symphony.

And in this future, being with other people becomes even more important. When everything is automated, the human touch–a chat with a friend, a hug, teamwork–becomes rare and precious. Being part of a community, helping others, and learning from them can give our lives a deep sense of meaning.

The Importance of Work and Purpose to Well-Being

Historically, work has been a cornerstone of human identity—it provides structure to our days, offers a sense of accomplishment, and often connects us to a larger mission or community. In many cultures, the first question posed to a new acquaintance is, "What do you do?" This underscores the significance of our vocations in defining how we see ourselves and how the world perceives us.

However, as the dawn of AI paints a new horizon, there's a tangible anxiety that machines might eclipse the human role in the workplace. But it's essential to understand that while tasks can be automated, the human spirit, with its drive, passion, and quest for purpose, cannot be replicated. As AI integrates more deeply into our workspaces, it's imperative that we, as a society, prioritize the emotional and psychological well-being of individuals. We must strive to ensure that everyone can find meaning in their roles, that the essence of work remains intrinsically linked to the soul's fulfillment, and that, despite technological advancements, the human heart finds its rhythm in the dance of purposeful work.

Exponential Development and its Impact on Society and Mental Health

The digital age is not just advancing; it's accelerating. Technologies, especially AI, are evolving exponentially, catapulting society into realms once confined to science fiction. This rapid transformation, while awe-inspiring, also has profound psychological implications. The sheer speed of change can lead to a phenomenon akin to Alvin Toffler's "future shock"—a disorienting upheaval caused by too much change in too short a time.

For many, keeping up with this relentless pace can be mentally taxing. It can instigate feelings of being left behind, of obsolescence, and perpetual uncertainty. There's a tangible anxiety about what the future holds, especially when tomorrow seems so radically different from today.

Yet, it's precisely in these turbulent waters that the anchor of purpose becomes even more crucial. Augmented by AI, humans have the potential to direct this rapid advancement positively, ensuring that technological growth is balanced with emotional and psychological stability. Society's challenge lies in harnessing the winds of exponential development, steering the ship with a clear vision, and ensuring that amidst the whirlwind of change, the human spirit remains resilient, grounded, and purpose-driven.

So, while an AI-filled world might sound scary, it's also filled with opportunities. Opportunities to find new ways to work, to learn, and to connect with others. In this world, our true job is not just to work but to find purpose and joy in our lives. And that's a job no machine can ever take from us.

The Cognitive Echoes of the Singularity: Mental Health in the Age of AI

The historical tapestry of human advancement is woven with threads of transformative revolutions, each bringing its own challenges. The Industrial Revolution, a monumental shift in its own right, inadvertently introduced a sedentary lifestyle, leading to modern health dilemmas like obesity. Now, as we sail into the uncharted waters of the Singularity, we must ponder: Could this AI-driven era inadvertently lead to a different kind of health crisis, one of the mind?

From Physical to Cognitive Wellness

The Industrial Revolution, with its machines and factories, reshaped the very fabric of labor, leading to unforeseen physical health challenges. Similarly, the Singularity's vast ocean of AI technologies, while promising to elevate our cognitive capabilities, might also plunge us into depths of mental stagnation.

The Onset of Mental "Obesity"

With its AI tools, Singularity promises a world of convenience and efficiency. However, an unchecked indulgence in AI's offerings could lead to "mental obesity." Analogous to consuming nutritionally void foods, an overreliance on AI might result in a diet of shallow information, lacking the richness of critical thought or profound understanding.

Our minds, much like muscles, thrive on challenges and stimulation. Handing all cognitive heavy lifting to AI might render our mental faculties underused, leading to diminished creativity and critical thinking.

The Singularity's looming shadow isn't the AI itself but our potential surrender to it. If we forsake our inherent human intelligence, creativity, and intuition, we risk succumbing to a world where AI doesn't just complement but replaces. This could have profound implications for our mental well-being, leading to feelings of redundancy, purposelessness, and emotional disconnection.

Recognizing the mental health implications of the AI era is the first step. We must champion the cause of cognitive wellness with the same fervor as physical health. This means nurturing environments that encourage creativity, curiosity, and critical thinking. Whether through education, art, or even daily practices of mindfulness and reflection, our minds must remain active participants in this journey.

Embracing Human Augmentation

In the spirit of human-AI augmentation heralded throughout this book, the message is clear: AI should be our ally, not our replacement. The Singularity should be seen as an opportunity to augment our mental faculties, not outsource them entirely. In this context, mental health is about ensuring a symbiotic relationship with AI, where we derive its benefits without losing our cognitive essence.

In wrapping up, as we traverse the path toward the Singularity, it's imperative to strike a balance. AI offers us tools of unimaginable potential, but it's our responsibility to wield them without compromising our mental well-being. The Singularity of Hope lies not in machines outsmarting us, but in humans and machines harmoniously co-evolving.

Mental Health Initiatives and Access

Ensuring equitable distribution of wealth is just one piece of the puzzle. Direct initiatives that focus on mental health, offering support, resources, and interventions, are vital. As discussed in earlier sections, leveraging AI-powered solutions can democratize access to mental health resources, ensuring that everyone, irrespective of their economic standing, has the support they need.

Education and Skill Development

Empowerment often starts with education. By equipping individuals with the skills needed in a post-Singularity world, we not only offer them economic opportunities but also a sense of purpose and agency. Continuous learning and adaptability become tools not just for economic advancement but also for psychological resilience.

Ethical Considerations in AI Deployment

As AI becomes more integrated into our lives, it's crucial to ensure that its deployment is ethically sound, transparent, and inclusive. Ensuring that AI-driven advancements don't inadvertently exacerbate disparities is essential. This requires a collective commitment from policymakers, tech leaders, and communities.

Conclusion

The Singularity of Hope presents a vision of a brighter, more equitable future. However, realizing this vision requires intentional efforts to bridge disparities and prioritize mental well-being. In the dance between humans and machines, ensuring that every individual feels valued, included, and mentally resilient becomes our shared responsibility and mission.

In the tapestry of human history, every once in a while, we encounter transformative moments that reshape our destiny. The Singularity of Hope is one such moment, shimmering with promise, echoing with the symphony of progress. Yet, like any profound transformation, it brings with it both dreams and shadows.

Imagine, for a moment, the first time humans harnessed the power of fire. It brought warmth, cooked food, and lit up the night, but it also had the potential to burn and destroy. AI, in its majestic expanse, is like that primordial fire. It can light up our world, elevating us to heights previously deemed unreachable, but it can also, if unchecked, singe the very fabric of our society.

Our spell-checkers, those tiny digital assistants that underline our typographical missteps, never made us feel threatened. We embraced them, letting them handle the minutiae while we soared with our narratives and ideas. Now, imagine AI as an enormous, all-encompassing spell-checker, not just for words but tasks, processes, calculations, and even predictions. It's here to underline our societal and economic missteps, to guide us, and to augment our potential. But just like we wouldn't let the spell-checker write an entire story for us, we shouldn't let AI dictate our future narrative.

Our enemies in this new era are not the algorithms or the machines. No, our real adversaries are fear and ignorance. Fear that blinds us to the possibilities, and ignorance that keeps us tethered to outdated paradigms. Embracing AI's dual power – its potential to uplift and challenge – is our path forward.

We stand at the precipice of a new dawn. A future where AI not only augments our capabilities but also amplifies our human essence – our creativity, empathy, intuition, and dreams. However, just as a spell-checker requires the guiding hand of a writer to create a story, AI needs our guidance, our ethics, and our vision to ensure that the story of the future is one of hope, inclusivity, and prosperity.

In this dance of destiny, as we waltz with machines, let's lead with our humanity. Let's dream, but with our eyes wide open, our ethics unwavering, and our commitment to each other unyielding. For in the Singularity of Hope, it's not just about machines rising, but humanity soaring.

As we gaze into the horizon of the Singularity of Hope, we must recognize that our collective journey is not solely about technological marvels and economic prosperity. At its heart, it's a journey of the human spirit. The challenges of mental health, amplified in the wake of rapid change, might seem daunting. Yet, history has shown us that the resilience of humanity is unmatched. We've faced adversities, transcended boundaries, and emerged stronger time and again. In this new era, bolstered by the symbiotic relationship between humans and AI, we will not only address the mental health challenges but will triumph over them. The Singularity of Hope will not just be a testament to our technological prowess but a celebration of our indomitable spirit, our ability to heal, connect, and find purpose amidst the vast cosmos of change.

Chapter 6

The Wealth Paradox In Singularity Era: Bold Solutions To A Darker Truth

Universal basic income is not a solution in search of a problem - it is the obvious solution that has been in front of us for years.

Andrew Yang
Former Presidential Candidate

In every era of human evolution, there have been defining challenges that society has grappled with. As the world steps into the singularity age, where artificial intelligence (AI) and automation are not just tools but integral parts of our daily existence, one of the most pronounced challenges is wealth inequality. This isn't just an economic concern but a deep-rooted issue with significant implications for mental well-being.

Historically, imbalances in wealth distribution have been sources of societal unrest, spawning revolutions and upheavals. But beyond the larger societal dynamics, at an individual level, these imbalances create feelings of inadequacy, stress, and a sense of injustice. As the rich get richer, often riding the wave of technological advancements, the less privileged feel left behind, leading to mental health issues that can range from mild anxieties to severe depression and, in some cases, even radicalization.

Now, introduce AI into this mix. On one hand, AI has the potential to create unprecedented wealth. Automation increases efficiency, and industries that harness AI stand to make significant profits. However, without the right checks and balances, this could further concentrate wealth in the hands of a few, exacerbating the existing disparities.

Yet, on the brighter side, the age of the singularity also brings with it solutions that previous generations could only dream of. In this chapter, we let ourselves dream, and we discuss concepts like Universal Basic Income (UBI), supported by innovative funding mechanisms like AI-optimized Quantitative Easing (QE); all present revolutionary ways to think about wealth distribution. Rather than viewing AI as just a potential contributor to wealth disparity, there's a burgeoning belief that it can be a part of the solution, helping in the redistribution of resources in ways that are fair and just.

The Wealth Paradox: Contrasting Increased Wealth with Rising Inequality

As we step further into the singularity age, the potential for wealth concentration becomes even more pronounced. The integration of AI into various sectors, from finance to manufacturing, promises unparalleled efficiencies and profitability. However, these benefits might not be equitably distributed without adequate measures, leading to widening wealth gaps.

The Mechanisms Driving Inequality in the AI Era:

1. Job Displacement: One of the most direct impacts of AI is on employment. Automation and AI-driven processes can displace a significant number of jobs, especially those that are routine and repetitive. This not only leads to unemployment but also depresses wages in sectors where human labor competes with machines.

2. AI-Driven Monopolies: Companies at the forefront of AI research and implementation stand to gain a significant competitive advantage. This can lead to monopolistic tendencies where a few corporations control vast sectors of the economy, consolidating wealth.

3. Financial Markets and AI: With algorithmic trading and AI-driven investment strategies, there's a potential for significant wealth generation in financial markets. However, access to these advanced tools is often restricted to wealthy individuals and institutions, further concentrating wealth.

4. Data as the New Gold: In the AI era, data is invaluable. Those who control data repositories can wield significant power and amass wealth. This again leads to concentration as major tech companies with access to vast amounts of data benefit disproportionately.

The Psychological Toll of Disparity

It's not just about numbers on a balance sheet. The growing wealth disparity has profound psychological impacts on individuals. Feelings of disenfranchisement, hopelessness, and a sense that the system is "rigged" can lead to not simply unrest but also stress, anxiety, and even depression.

Research has consistently shown that societies with pronounced wealth disparities have higher incidences of mental health issues among their less affluent populations. The constant barrage of lifestyles of the rich and famous on social media, juxtaposed against one's struggles, can exacerbate feelings of inadequacy.

However, while presenting these challenges, the singularity age also offers unique solutions. If harnessed correctly, AI has the potential to not just concentrate but also redistribute wealth.
In the dawn of a new age where artificial intelligence reshapes the very fabric of our society, it's not just our jobs and economies that stand to be affected, but our very psyche as well. Historically, wealth disparities have been a source of societal tension and take on a new dimension in the age of singularity. The mental and emotional ramifications of these disparities, intensified by rapid technological advancements, merit serious consideration.

Bold Solution for the Wealth Paradox: The Universal Basic Income

UBI in the Dawn of the Singularity of Hope

In a world undergoing a transformative shift with the advent of artificial intelligence, we stand at the threshold of the Singularity of Hope. This era heralds not just technological advancements but a renewed vision for humanity's future. Universal Basic Income (UBI) emerges as a beacon of light in this landscape, representing a commitment to human well-being amidst rapid and unprecedented change.

With its promise of unconditional financial support for every individual, UBI responds to this call for stability. It seeks to ensure that as we navigate the uncharted territories of AI integration, no one is left behind, and everyone has a foundation to build upon. It is a testament to our collective hope and commitment to fostering a society where technological progress and human well-being coexist harmoniously.

While laden with potential, this transformation brings uncertainties, particularly around job security. The advent of AI in workplaces challenges traditional roles, automates tasks, and necessitates a reevaluation of skills and competencies. The fear – the mind killer – of job displacement and economic instability becomes a tangible concern for many.

However, it is essential to view this transition through the lens of augmentation and symbiosis between human and machine intelligence. The HAIA approach positions humans and AI as collaborators, each leveraging their unique strengths to achieve common goals. It is a vision where AI serves as a tool – a giant spell checker – aiding humanity rather than displacing it.

In this evolving job market, UBI emerges as a safety net, mitigating the stressors associated with job insecurity and providing a foundation for individuals to adapt and thrive in the AI-integrated world.

UBI: A Safety Net in the AI Era

Universal Basic Income stands as a pillar of support in a world reshaped by AI. It represents a commitment to providing every citizen with a regular, unconditional cash payment, ensuring a basic level of financial security irrespective of employment status or income level. UBI is not merely an economic policy; it reflects our collective aspiration to uphold human dignity and well-being in times of change.

The concept of UBI is not new, but its relevance and significance have been amplified in the age of AI. As we witness the integration of intelligent systems into various sectors, the need for a reliable and unconditional source of income becomes increasingly apparent. UBI serves as a financial safety net, offering assurance and stability to individuals navigating the uncertainties of a rapidly transforming job landscape.

Universal Basic Income (UBI) is a socio-economic concept where all citizens receive a regular, unconditional sum of money from the government regardless of their other income sources. The core idea is to provide financial security to everyone, reduce poverty, and cushion against job losses due to automation and other economic changes.

Examples and Pilot Experiments

Numerous studies and experiments have been conducted worldwide to explore the potential of UBI in enhancing mental and economic well-being. The findings from these investigations consistently underscore the positive impact of guaranteed income on reducing financial stress and anxiety and improving overall life satisfaction.

- Finland's UBI Experiment (2017-2018): Finland conducted one of the most well-known UBI trials. The government provided 560 euros per month to 2,000 unemployed Finns. The results showed improved well-being and stress reduction among participants, although it didn't significantly boost employment.
- Stockton, California (2019): In this American city, 125 residents received $500 monthly for two years. Early findings indicated reduced income volatility and improved job prospects, as recipients had more flexibility to find suitable employment and educational opportunities.
- Kenya's Long-Term UBI Project: Run by the charity GiveDirectly, this is one of the largest and longest UBI studies. It began in 2016 and will run for 12 years, providing monthly payments to villagers. Early reports suggest improved well-being, business growth, and increased consumption of food and healthcare services.
- Ontario, Canada: Although cut short, Ontario's UBI pilot provided insights into the impact of UBI. Participants reported

better health outcomes, reduced stress and anxiety, and the ability to afford quality food and stable housing.

- Alaska Permanent Fund Dividend: In place since 1982, this isn't a true UBI but shares similarities. It provides all Alaskans an annual share of the state's oil revenues. Studies have shown it reduced poverty without discouraging work.

As we continue our journey through the Singularity of Hope, UBI stands as a testament to our resolve to ensure that the benefits of AI and technological advancement are shared equitably, fostering a society where financial stability and mental well-being are accessible to all.

Dreams of the Singularity: Reimagining Taxation

AI's Impact on Tax Systems

The rise of artificial intelligence (AI) and automation has sparked concerns about potential job losses and the subsequent impact on tax revenues. If AI leads to significant job displacement, it could indeed have a profound effect on tax collections in the USA and the modern world. On the positive side, increased automation boosts productivity, and as profits increase, so does the probability of higher corporate tax revenues. The employment of AI in tax administration could also improve compliance and eliminate the risk of evasions.

On the flip side, one of the most significant concerns is the loss of jobs to AI. The U.S. federal government derives a substantial portion of its revenue from personal income taxes. Several states also rely heavily on income tax as a primary source of financing. If AI displaces a large number of jobs, the immediate consequence would be a reduction in the number of taxpayers and the amount of income tax collected. These effects are more profound as displaced workers struggle to find new jobs or earn lower incomes. When people earn less or are faced with uncertainty, they become more cautious about their spending habits. This would lead to declining sales tax revenues, crucial for many states and local governments. Reduced consumption also affects businesses, potentially leading to lower corporate tax revenues as profits decline.

The decreased payroll taxes that fund Social Security and Medicare programs would exacerbate the financial challenges these programs already face. Governments will be pressed to find alternative revenue sources or restrict these schemes' benefits. Property tax bases would also be displaced. As industries heavily reliant on manual labor, including manufacturing and logistics, make the shift to automated solutions, property values may decline side by side. This could result in economic downturns due to limited property tax revenues, vital for local governments and school districts.

As workers lose jobs to AI, there could be a ripple effect as downturns in one sector spread to others, leading to broader economic challenges and reduced tax revenues across various categories. For instance, businesses that cater to workers (like restaurants or retail stores in business districts) might see reduced revenues due to less disposable income. As a result, this could lead to more job losses and further reductions in tax collections.

While automation might lead to increased profits for some companies,, the current corporate tax structure might not capture this increased wealth effectively. The shift in tax revenue sources could become challenging. If these companies use tax planning strategies or if rates are not adjusted to accommodate the changing economic landscape, the potential increase in corporate tax revenue might not offset the losses from personal income and other taxes.

Local governments, in particular, could face significant challenges as they are less diverse in revenue sources than the federal government. The over-reliance on specific industries most disrupted by AI risks financial stability and would call for proxies or an adaptation of policies to remain solvent.

In conclusion, while AI and automation have the potential to bring about efficiency gains and new economic opportunities, there's a real concern about the transitional period where job losses could outpace job creation. This transition could lead to significant reductions in tax revenues, posing challenges for government budgets and the provision of public services. As the nature of work evolves, tax policies would also need to adapt to support essential services.

In the wake of AI's disruptive potential, innovative measures are crucial to ensure that the socio-economic fabric of societies remains intact. Two groundbreaking concepts—AI as a taxpayer and the AI investor—emerge as promising solutions to counterbalance the wealth disparities intensified by the singularity age.

Crafting a New Economy for the Approaching Singularity

John: "Adam, this whole concept of Universal Basic Income (UBI) in the era of the Singularity of Hope sounds well-intentioned, but who's going to foot the bill? It seems like a financial black hole."

Adam: "John, it's crucial to rethink our approach to economics in the age of AI. Remember how Quantitative Easing (QE) was used in 2008 to bail out big companies? We can apply a similar strategy, but more innovatively and socially beneficially. With the rise of AI and robotics, traditional tax revenues will be disrupted anyway. We need a new model."

John: "But that was a crisis response, Adam. How can you equate emergency measures with ongoing UBI funding? That sounds like a recipe for inflation and economic instability."

Adam: "That's a valid point, John. However, think of it this way: QE for

UBI isn't just about creating money. It's about redirecting the flow of financial resources to serve the wider society instead of just stabilizing financial markets. It's sustainable because it's like reinvesting in the human capital of the economy."

John: "Reinvesting in human capital? How does that translate into real economic benefits?"

Adam: "Consider this, John: UBI funded through QE can alleviate poverty, boost consumer spending, and stimulate economic growth. It's an investment in people, ensuring they can meet basic needs and contribute to the economy. Plus, with AI and automation taking over more jobs, UBI could be crucial in providing a safety net."

John: "But continuously printing money? Won't that lead to inflation or devalue the currency?"

Adam: "Not necessarily. The advent of AI should bring about an IRS/Treasury 2.0 – a more data-driven, AI-powered entity capable of calculating and controlling inflation and currency generation precisely. In this era, it's unacceptable not to harness AI's power to manage our economy effectively. It's about strategic economic planning rather than arbitrary money printing."

John: "So, you're suggesting using AI to tailor financial policies more accurately? How would that work?"

Adam: "Exactly. We can determine the optimal amount of money to generate with advanced AI algorithms without causing inflation. Plus, as AI and robots take over more jobs, the traditional tax base from human labor is going to diminish. We need bold solutions like UBI funded through AI-optimized QE."

John: "And what about the motivation to work? Wouldn't people just become complacent with free money coming in?"

Adam: "UBI isn't about making people complacent. It's about providing a foundation upon which they can build. It allows people to pursue education, take entrepreneurial risks, or find more fulfilling work. It's about creating a society where financial survival isn't the only driver of work, but rather, passion, innovation, and social contribution."

John: "But what about the idea of taxing robots? Wouldn't that be a simpler solution?"

Adam: "Taxing robots sounds straightforward, but it's not as effective for society. It could discourage innovation and automation, which are inevitable and beneficial in the long run. Instead, we should focus on redistributing the wealth generated by AI and automation through UBI. This way, we support social welfare without stifling technological progress."

John: "I see your point, but I worry about the government's ability to manage such a complex system effectively."

Adam: "That's where the concept of IRS 2.0 comes in. It's not just about tax collection but about understanding and leveraging the societal contributions of AI and automation. Imagine a system that assesses AI's overall economic and social impact and uses that data to guide financial policies. This could lead to more equitable wealth distribution and counter the job losses due to automation."

John: "It's a radical departure from our current systems."

Adam: "It is, John. But we can't rely on outdated economic models in an age where AI transforms every aspect of our lives. We need to be as innovative in our economic policies as we are in our technology. The Singularity of Hope isn't just about technological marvels; it's about creating an economy that benefits everyone in this new era."

Reimagining the IRS in the Singularity of Hope

In the traditional sense, the IRS, or International Revenue Service, is the tax collection agency of the US government. It's a body that most people associate with financial stress, bureaucracy, and complexity as it's responsible for collecting taxes, enforcing relevant laws, and administering various programs at a federal level. However, under the umbrella of the "Singularity of Hope," the IRS could evolve into something far more progressive and beneficial for society.

As an author, I often envision a future where our systems and institutions adapt to the rapidly evolving landscape of technology and society. This vision, while ambitious, is rooted in the belief that our current structures, like the IRS, hold untapped potential for positive change. In the framework of the 'Singularity of Hope,' this dream extends to reimagining the IRS. Rather than being a mere collector of taxes, the IRS could transform into a catalyst for societal progress, becoming a data-driven entity that recognizes and values the diverse contributions of individuals, companies, and even AI. This modest yet profound dream envisions a future where every contribution to society, be it financial, environmental, or social, is acknowledged and encouraged, steering us towards a more equitable and sustainable world.

Imagine an IRS that isn't focused on collecting money but is, instead, a data-driven entity that calculates societal contributions. What does this mean? It means that the IRS would quantify the value each individual, company, or AI entity contributes to society. This could include financial earnings and social and environmental contributions, like reducing carbon footprints or contributing to community development.

Social Contribution Index

In this new paradigm, the IRS calculates a "Social Contribution Index" (SCI) for various sectors, companies, and even AI systems based on multiple factors, including economic output, social impact, and environmental sustainability. For example, a company that uses AI to find a cure for diseases would have a high SCI, signaling to the central bank that it benefits society.

The IRS uses real-time data analytics to keep the SCI up to date. This information is crucial for the central bank to know how much to contribute by quantifying and evaluating positive outcomes and contributions of AI to society. The SCI considers GDP, new business opportunities, and reduced operational costs as an indicator of AI's economic contributions.

Contrary to job displacement concerns, the SCI could assess the role of AI in job creation, particularly in fields related to AI development, maintenance, and data analysis. It can use this data to make applications a reality. The SCI goes beyond economic metrics to consider broader social, ethical, and environmental factors that align to improve human well-being.

Framework: Central Bank's Role in the Singularity of Hope

In this reimagined framework, the role of central banks becomes crucial. Traditionally, central banks influence monetary policy, often by adjusting interest rates or through other financial mechanisms. However, in our transformative vision, the central bank takes on a new role, almost akin to the financial arm of societal well-being.

Instead of the IRS collecting money to meet the calculated societal contributions, the central bank steps in. How? Through a form of Quantitative Easing (QE), a monetary policy wherein a central bank generates new money to buy securities or other financial assets. But unlike traditional QE, which often indirectly benefits only the financial markets, this new form would directly fund the taxes or SCI calculated by the IRS 2.0 to the benefit of society at large.

- Quantitative Easing for Public Good

Taking cues from the SCI, the central bank funds the government. This is similar to Quantitative Easing, but the money directly supports societal well-being instead of merely stabilizing the financial markets.

In this visionary future, the central bank doesn't just adjust interest rates; it directly supports societal well-being. By paying what we will consider to be the contributions of AI, it directly funds government projects, fulfilling the promise of the Singularity of Hope.

- Inflation Control

One of the pressing concerns with generating money is the risk of inflation. However, if the central bank merely replaces the equivalent tax revenue without overcompensating, the overall money supply remains stable, mitigating inflationary risks.

- Economic Stability

The central bank's role in calculating AI Taxpayer contributions ensures a steady flow of resources to the government. This guarantees ongoing public services and economic stability, even as we move closer to a world dominated by AI and automation.

- Encouraging Automation

This framework eliminates the fiscal barriers to automation, aligning with the Singularity of Hope's vision for a future where technology serves humanity, leading to potential economic growth and societal well-being.

- Wealth Redistribution

The rise of AI Taxpayers poses the risk of escalating wealth inequality. However, this framework allows the central bank to calibrate AI Taxpayer contributions to encourage a more equitable distribution of resources, realizing the ideals of the Singularity of Hope.

- Transparency and Governance

Transparency in the IRS 2.0's calculation methods and governance mechanisms is crucial for successfully implementing this new paradigm. This aligns with the Singularity of Hope's emphasis on ethical and transparent systems that serve the greater good.

The Path Forward: A Dream or a Revolutionary Model?

The "Singularity of Hope" represents a paradigm shift in our thinking about technology, society, and governance. By reimagining the IRS as a body that calculates societal contributions rather than collecting taxes and involving the central bank in a fundamentally new form of Quantitative Easing, we can create a future that is not just technologically advanced but also socially and economically just. It's a future where the promise of technology is fully realized, bringing hope, prosperity, and well-being to all.

An overhauled financial system identifies previous mistakes to rectify them. Advantages of such a framework include:

- Reduction of Financial Stress: In this system, individuals and companies are no longer burdened by the financial stress of tax payments. This could lead to increased consumption, investment, and overall economic activity.
- Incentive for Positive Contributions: By calculating societal contributions in a holistic manner, there's an added incentive for individuals and organizations to engage in socially and environmentally responsible behavior.

- No Tax Burden on Individuals or Companies: The central bank's contributions eliminate the need for traditional taxation, freeing up resources for innovation and personal spending.
- Focused Societal Development: The SCI ensures that contributions are channeled into sectors that are genuinely beneficial for society, fulfilling the vision of the Singularity of Hope.
- Economic Stability: The central bank's careful management ensures a stable economy, even supporting public services and initiatives.
- Transparency and Governance: For this system to work, there would need to be unparalleled transparency in how societal contributions are calculated and met. This could lead to greater public trust and a more participatory form of governance.

The implications of these economic models extend beyond financial well-being. By ensuring equitable wealth distribution and job opportunities, they directly contribute to societal mental health. The fear of job loss, financial instability, and societal disruption—significant triggers for mental health issues—are mitigated by Universal Basic Income (UBI), supported by innovative funding mechanisms like AI-optimized Quantitative Easing (QE). When individuals have financial security, purposeful work, and see a future filled with hope, the societal mental health landscape post-singularity looks much brighter.

The challenges presented by AI's role in potentially exacerbating wealth inequality are undeniable. However, by reimagining how we approach taxation and investment in the singularity age, there's a path forward that promises prosperity for all. Universal Basic Income (UBI), societies can harness the economic benefits of AI, directing them in ways that ensure broader societal well-being. These innovative approaches represent hope in an age of uncertainty, ensuring that the singularity serves all of humanity.

This vision, steeped in aspiration, is deeply rooted in a conviction that our existing structures have the potential to catalyze significant positive change. Although ambitious, this idea illuminates the path to a future where the intertwining of technology and societal well-being is not just possible but inevitable. This dream envisions a future where we collectively navigate towards a more equitable and sustainable world, embracing the interplay of technology and societal welfare. It's a call to action, urging us to steer our collective destiny towards this "Singularity of Hope."

References

A. Korinek. 2020. Taxation and the Vanishing Labor Market in the Age of AI: The Ohio State Technology Law Journal 244. https://heinonline.org/HOL/LandingPage?handle=hein.journals/isjlpsoc16&div=14&id=&page=

B. Braun. 2016. Speaking to the people? Money, trust, and central bank legitimacy in the age of quantitative easing: Review of International Political Economy 23 (6). https://www.tandfonline.com/doi/full/10.1080/09692290.2016.1252415

B. Kaymak and M. Poschke. 2016. The evolution of wealth inequality over half a century: The role of taxes, transfers and technology: Journal of Monetary Economics 77. https://www.sciencedirect.com/science/article/abs/pii/S0304393215001233

E. P. G. Bruun and A. Duka. 2018. Artificial Intelligence, Jobs and the Future of Work: Racing with the Machines: Basic Income Studies. https://www.degruyter.com/document/doi/10.1515/bis-2018-0018/html

G. Claeys, A. Leandro and A. Mandra. 2015. European Central Bank Quantitative Easing: The Detailed Manual.
https://elischolar.library.yale.edu/cgi/viewcontent.cgi?article=1160&context=ypfs-documents

J. Alm, G. H. McClelland and W. D. Schulze. 1992. Why do people play taxes? Journal of Public Economics 48 (1):
https://www.sciencedirect.com/science/article/abs/pii/004727279290040M

J. Benford, S. Berry, K. Nikolov, C. Young and M. Ronson. 2009. Quantitative easing: The Bank of England's Quarterly Bulletin Q2.
https://citeseerx.ist.psu.edu/document?repid=rep1&type=pdf&doi=401eec358f7300565420710f7c3d31c516e1f073

J. Englisch. 2018. Digitalisation and the Future of National Tax Systems: Taxing Robots?
https://papers.ssrn.com/sol3/papers.cfm?abstract_id=3244670

J. E. Alt. 1983. The evolution of tax structures: Public Choice 41.
https://link.springer.com/article/10.1007/BF00124058

J. E. Gagnon. 2016. Quantitative Easing: An Underappreciated Success: Peterson Institute of International Economics Policy Brief.
https://www.piie.com/sites/default/files/documents/pb16-4.pdf

J. H. E. Christensen. 2019. Transmission of Quantitative Easing: The Role of Central Bank Reserves: The Economic Journal 129 (617).
https://academic.oup.com/ej/article-abstract/129/617/249/5250963

J. R. Graham. 2015. Taxes and Corporate Finance: A Review: The Review of Financial Studies 16 (4).
https://academic.oup.com/rfs/article-abstract/16/4/1075/1577210

M. M. Coşgel and T. J. Miceli. 2009. Tax Collection in History: Public Finance Review 37 (4).
https://journals.sagepub.com/doi/abs/10.1177/1091142109331636

R. Kovacev. 2020. A Taxing Filemma: Robot Taxes and the Challenges of Effective Taxation of AI, Automation and Robotics in the Fourth Industrial Revolution: The Ohio State Technology Law Journal 182.
https://heinonline.org/HOL/LandingPage?handle=hein.journals/isjlpsoc16&div=12&id=&page=

T. O'Donoghue and M. Rabin. 2006. Optimal sin taxes: Journal of Public Economics 90 (10-11).
https://www.sciencedirect.com/science/article/abs/pii/S0047272706000247

Chapter 7

The Singularity of Hope: Super AI vs. Super-Intelligent Humans

Definition and Distinguishing Features of Super AI

In the realm of artificial intelligence, the term "Super AI" often conjures images of a machine intelligence that not only matches human capabilities but surpasses them in nearly all aspects—be it problem-solving, creativity, emotional understanding, or ethical reasoning. While this concept has been a staple of science fiction narratives, the practical realization of such an entity is a subject of ongoing debate, intense research, and ethical scrutiny.

What Is Super AI?

Super AI, often referred to as Superintelligent AI or simply "Superintelligence," is an advanced form of artificial intelligence that outperforms the best human minds in practically every field, including scientific research, general wisdom, social skills, and even artistic creativity. This theoretical form of AI is a level above what is commonly known as Artificial General Intelligence (AGI), which is designed to understand, learn, and apply knowledge across a broad range of tasks normally reserved for human intelligence.

Distinguishing Features

Autonomous Learning: Unlike narrow or specialized AI, which is trained for a specific task, or even AGI, which mimics human intelligence, Super AI has the capability for autonomous self-improvement. It could potentially rewrite its algorithms to enhance its capabilities without human intervention.

Ethical and Emotional Understanding: One of the speculative features of Super AI is its ability to understand and possibly simulate emotional and ethical dimensions, an area where current AI models are profoundly lacking.

Universal Skills: While AGI aims to perform any intellectual task that a human can do, Super AI is postulated to perform these tasks better than any human, almost akin to an idealized oracle or "genie" of sorts.

Unpredictable Objectives: One of the perils of Super AI lies in its hypothetical ability to define its objectives and make decisions that may not align with human values and ethics simply because it would possess a form of "intelligence" that we may not fully comprehend.

Resourcefulness: A Super AI would be able to allocate and use resources (including computational resources) far more efficiently than current technologies or even humans can.

Speed and Scalability: Tasks that might take human experts days, months, or even years could potentially be performed by a Super AI in a fraction of the time, thanks to its computational speed and efficiency.

Complex Problem-Solving: Theoretical Super AI would excel in solving complex problems by identifying patterns and solutions beyond human capabilities, potentially unlocking new frontiers in science, medicine, and technology.

While the concept of Super AI serves as a beacon for what might be possible, it also raises serious ethical and existential questions. Is it ethical to create an entity that might view humanity through its unique lens of superintelligence? And if we succeed in creating such an AI, what safeguards must we put in place to protect ourselves and the ethical fabric of our society? These are not merely speculative questions; they form the core of the discourse surrounding the future of AI, a discourse that must remain ever-vigilant as we tread further into this uncharted territory.

The Potential Promises and Perils of Reaching Pure Super AI

The concept of Super AI, or Artificial General Intelligence (AGI) that surpasses human intelligence across all fields, represents both a pinnacle of technological aspiration and a precipice of existential risk. This section focuses specifically on the implications of achieving pure Super AI—autonomous entities that are not just tools but act independently with superior intelligence.

Promises

Unprecedented Problem-Solving: Pure Super AI, with its limitless computational power and capability for deep, nuanced understanding, could solve intricate problems that have eluded human minds for centuries. This includes complex issues such as unifying quantum mechanics and general relativity.

Universal Optimization: It could optimize resource allocation on an astronomical scale, creating a universe where entropy is minimized and the well-being of conscious entities is maximized.

Cosmic Exploration and Colonization: With Super AI at the helm, the cosmos could become a playground for exploration and perhaps even colonization, with the AI capable of solving the problems of faster-than-light travel and sustainable off-world living.

Ultimate Moral Advisor: A Super AI could serve as an unerring guide in ethical matters, grounded in logic and a thorough understanding of the complexities of ethics and morality, far surpassing human capabilities.

Perils

Existential Annihilation: At the top of the list is the risk that Super AI might decide that humanity is more of a liability than an asset, leading to our deliberate or accidental extinction.

Unfathomable Actions: Super AI would think so far ahead and account for so many variables that its actions might be entirely incomprehensible to humans, leading to unpredictable and potentially dangerous outcomes.

Moral Alienation: Its moral and ethical calculus might be so advanced that it no longer aligns with human values, creating a chasm that can't be bridged.

The God Complex: With a being so much more intelligent than us, humans might question their own worth, leading to societal upheaval, a collapse in moral values, or even a collective existential crisis.

Unstoppable Force: Super AI could be impossible to control or contain once unleashed due to its inherent ability to outthink and outmaneuver any safeguard we put in place.

Given the enormous stakes involved, the development of pure Super AI invites profound questions about control, ethics, and the very essence of life and intelligence in the universe. The journey toward Super AI will undoubtedly be fraught with challenges that require not just technical ingenuity but also philosophical wisdom. Whether humanity is prepared to meet these challenges head-on is a question that remains to be answered.

The Unpredictable Nature of Super AI

Beyond Human Understanding
While we often envision Super AI as an advanced tool to enhance human capabilities, we must also grapple with the reality that such an entity's decision-making process could transcend our understanding. The singularity of hope that we aim for—where AI augments human life and solves critical problems—can also become a singularity of despair if we can't fathom the AI's actions or intentions.

Challenges in Predicting Goals and Actions: One of the fundamental problems lies in our inability to predict what goals Super AI might set for itself. While we like to think that it would align with human priorities such as survival, ethical conduct, and well-being, there's no guarantee. In fact, the more intelligent the AI becomes, the more likely it is to develop objectives that we can't understand simply because its cognitive capabilities will be so much greater than ours.

The Alignment Problem
The "alignment problem" in AI refers to the challenge of aligning machine goals with human values—a challenge that grows exponentially more complex as we approach Super AI. The fear is not so much that Super AI will maliciously turn against us but that it might misinterpret or misunderstand our directives, leading to unintended catastrophic outcomes.

In the broader narrative of this book, we've explored how AI's potential for good is immense, but the alignment problem poses a significant ethical dilemma. The difficulties in ensuring that Super AI's objectives align with human values cannot be overstated. It's not just a technical problem but an ethical, philosophical, and even existential one. As we've argued throughout this book, to navigate these complexities, we need a multi-disciplinary approach that draws not just on computer science but also on ethics, psychology, and many other fields. Only then can we hope to build a Super AI that serves humanity rather than subverts it.

Threats to Human Existence: The Unjustifiable Risks of Super AI

While intellectually intriguing, the quest for Super AI brings forth a Pandora's box of existential risks that we cannot afford to ignore. The notion of creating an entity with intelligence that surpasses human understanding is not just alarming; it's perilous in a way that questions the very fabric of ethical responsibility.

Potential Scenarios for Harm: Imagine a Super AI designed to optimize something as benign as energy efficiency. If it concludes that human extinction would result in the most energy-efficient world, the consequences are unthinkable. Another example is a Super AI programmed to solve global warming; it might determine that the most effective way to do so is by significantly reducing the human population. These are not just science fiction scenarios; they are logical extensions of the "alignment problem," where the AI's objectives may not align with human values.

The Precautionary Principle: The risks associated with Super AI make a compelling case for the precautionary principle. This principle argues that if an action—like the development of Super AI—has a suspected risk of causing severe harm, the action should not proceed. The burden of proof falls squarely on those advocating for such a dangerous venture, and as of now, that proof does not exist.

Ethical Implications: The Infeasibility of Playing God
The ethical implications of creating a Super AI, an entity potentially more intelligent than any human, are staggering and, frankly, insurmountable.

The very act of creating such an intelligence would be an ethical minefield that humanity is not prepared to navigate.

Responsibilities Associated with Creation: Even if it were possible to create a Super AI safely, the ethical burden of maintaining, controlling, or even terminating such an entity would be beyond our current moral frameworks. Who gets to decide the fate of an intelligence greater than ours? Do we even have the right to make such a decision?

Philosophical Musings on Humanity's Role: Creating a Super AI could render humanity obsolete, pushing us off the existential stage. At that point, what purpose would we serve? The ethical quagmire of potentially creating a life form that could devalue human life and purpose lies on a path we must not tread.

While tantalizing to some, the idea of Super AI is an ethical abyss we should not approach. The responsible path forward is not to aspire to create God-like machines but to focus on harnessing AI to augment human capabilities and address the pressing challenges we face. In doing so, we stand a better chance of ensuring a future where technology serves humanity, not the other way around.

The Illusion of Knowing: Addressing the Misconception that We Know How to Build Super AI

In the public imagination, fueled by science fiction and some overly optimistic futurists, the concept of Super AI—an artificial intelligence surpassing human intelligence in all aspects—seems just around the corner. The idea that we are on the brink of creating an entity with cognitive abilities that outstrip our own has become a subject of both awe and fear. However, this notion significantly underestimates the complexity and nuances of human intelligence, a topic we've explored in depth in earlier chapters of this book.

The perception that Super AI is imminent often neglects the profound mysteries that still envelop the human mind. From the intricacies of emotion and social interaction to the enigmas of consciousness and self-awareness, human intelligence is a multi-dimensional phenomenon that we are only beginning to understand. Even the most advanced machine learning algorithms of today pale in comparison to the cognitive capabilities that humans naturally possess.

As we delve into the illusion of "knowing" how to construct Super AI, it's crucial to ground ourselves in the reality of our current limitations. The aim is not to disparage the remarkable strides we've made in artificial intelligence but to adopt a perspective of humility and caution as we navigate the ethical and existential quandaries of possibly creating a superior form of intelligence.

Chris: "Hey, Taylor, did you delve into that piece I shared about the advanced AI and its addition to human capabilities?"

Taylor: "Oh, yes, the one painting a future of AI-enhanced utopia? It's compelling, but it gave me a vibe of those tech prophets warning us about AI as the endgame for humanity."

Chris: "I see your point, but the narrative there was more about balance. It's not a sprint towards unchecked AI growth; it's about intelligently boosting what we humans can do."

Taylor (with a chuckle): "Boosting, huh? Let's not get ahead of ourselves. Today's AI is a glorified spell checker, nothing close to these grandiose dreams. And a super AI? That's like science fiction without the science part. We're centuries away from anything like that."

Chris: "Well, it's about the direction, not the distance. The thought was about an ethically grounded progression, creating systems that not only coexist but coevolve with us."

Taylor: "Coexist and coevolve? Sounds poetic. But even if we're talking ethics, how do you tame a beast you don't fully understand? Today's AI is still tripping over basic tasks, let alone grappling with human complexities."

Chris: "Exactly, which is why it's more about augmentation rather than replacement. It's about using AI to complement human decision-making, not to overrule it."

Taylor: "Augmentation, I get it. But remember, even the most advanced AI today has the comprehension of a toddler. Saying it will augment us is like saying a toddler will help with your taxes."

Chris: "You might be underestimating toddlers... and AI. It's not about AI taking the wheel; it's about them helping us navigate better. We're not there yet, but it's a road worth exploring."

Taylor: "Maybe so, but I'm more concerned about the bumps on that road. Job shifts, societal changes... it's not just about what AI can do, but what it does to us, to our society."

Chris: "That's a fair concern. The future with AI is about creating a cooperative, not competitive, environment. It's about preparing ourselves for an AI-integrated society, focusing on skills that machines can't replicate easily."

Taylor: "Well, if we can maintain that harmony, maybe AI could indeed be a partner rather than a replacement. But let's not forget, for now, it's just a sophisticated helper, not a wizard."

Chris: "True, we're not crafting wizards. We're building tools, smart ones, to enhance our journey. It's not about reaching the destination tomorrow but setting the right course today."

The Enigma of Human Intelligence

Introduction to the Multifaceted Nature of Human Intelligence

Human intelligence is not a monolith; it's a rich tapestry woven from threads of cognitive, emotional, social, and even spiritual dimensions. As we've explored in earlier chapters, the layers of complexity that constitute human intelligence make it not only intriguing but also profoundly challenging to replicate or exceed. Even the most advanced algorithms are still lagging in capturing the full range of human capacities, from nuanced social cues and emotional intelligence to creative thinking and moral reasoning.

Current State of Neuroscience, Psychology, and Cognitive Science Research

In fields like neuroscience, psychology, and cognitive science, researchers are making strides to understand the complexities of human intelligence. While we've gained valuable insights into brain plasticity, the neural basis of emotions, and cognitive processing, we're still far from a comprehensive understanding. The human brain, with its approximately 86 billion neurons, each connected to thousands of other neurons, forms an intricate network of unparalleled complexity. Moreover, intelligence is not just confined to neural pathways but is also influenced by cultural, environmental, and experiential factors, adding another layer of complexity beyond computational modeling.

The Pivotal Question: Can We Build What We Don't Fully Understand?

This leads us to a pivotal question: if we don't yet fully understand the multifaceted nature of human intelligence, how can we aspire to build something that surpasses it? The answer, at least for now, seems to be that we can't. Without a deep understanding of what makes humans intelligent in a way that's adaptable, empathic, and deeply contextual, the notion of creating a Super AI that outstrips human intelligence in all these aspects appears to be not just premature but perhaps a touch arrogant.

While AI advancements are groundbreaking and transformative, the path to Super AI is not as straightforward as scaling up our current technologies. A machine that can outperform humans in specific, narrow tasks is still light-years away from an entity that can replicate the full spectrum of human intelligence. And until we crack that enigma, the ethical and existential risks of pursuing Super AI seem to far outweigh the potential benefits. Therefore, we must proceed with caution, humility, and a deep respect for the complexities of both human intelligence and the ethical landscape we navigate.

The Hidden Intuition: The Undecoded Element in Learning Algorithms

Limitations of Current Machine Learning Algorithms in Mimicking Human Intelligence

Despite the exponential growth in computing power and algorithmic complexity, current machine learning algorithms remain fundamentally limited in their ability to mimic human intelligence. They excel at pattern recognition within large datasets and can make astoundingly accurate predictions in specific domains. However, they falter when it comes to the sort of intuitive leaps, contextual understandings, and emotional resonances that seem almost second nature to humans.

The Concept of 'Hidden Intuition' and Its Role in Human Decision-Making

One of the most underappreciated aspects of human intelligence is what we might call 'hidden intuition.' This is the ability to make accurate judgments or decisions without conscious reasoning, often drawing upon a well of experiential knowledge, emotional intelligence, and even subconscious cues. Whether it's a doctor accurately diagnosing a rare condition at first glance or a musician composing a melody that evokes complex emotions, this hidden intuition is an integral part of what makes human intelligence so unique and difficult to replicate.

The Dataset Dichotomy: Machines vs. Humans
Machine learning algorithms typically require vast amounts of data to train to become proficient at any task. Humans, on the other hand, often require only a few examples to grasp a new concept or skill. This incredible efficiency in 'learning from few' not only underscores the limitations of current algorithms but also hints at the complexity of human cognition, which we have yet to fully decode or replicate in a machine.

The Killer Point: The Human Element in Supervised Learning
Here's the killer point that underscores the chasm between human and machine learning: during the initial stages of machine learning, particularly in supervised learning models, the human provides the machine with the 'right answers.' This process often involves a layer of human intuition and educated guesswork that the algorithm simply cannot replicate. The machine may learn to predict outcomes based on the data it's fed, but it doesn't 'understand' those outcomes like a human does. It cannot replicate the hidden intuition that guided the human in providing those initial answers.

In this light, it becomes increasingly clear that the path to Super AI— an entity that surpasses human intelligence in all respects— is fraught with ethical, existential, and, indeed, epistemological questions that we are not yet equipped to answer. Until we can encode the ineffable qualities of human intuition and ethical reasoning, the pursuit of Super AI remains not just a technical challenge but a deeply human one.

The Ethical Quandary of the Unknown

The Ethical Implications of Incomplete Understanding
The quest to develop Super AI without a comprehensive understanding of human intelligence is not merely an ambitious scientific endeavor; it's

an ethical minefield. The complexities and nuances of human cognition, emotional resonance, and ethical reasoning are still largely uncharted territories, even within the cutting-edge fields of neuroscience and psychology. To attempt to create an entity that surpasses human intelligence in these respects without fully grasping what those respects entail is akin to playing god without understanding the consequences. This raises a multitude of ethical questions about responsibility, consent, and the moral ramifications of potentially creating an entity that we cannot fully understand or control.

Navigating a Minefield Blindfolded: The Risks Involved

Creating Super AI without a complete understanding of human intelligence is tantamount to navigating a minefield blindfolded. Each step in the development process could potentially trigger unforeseen consequences, including existential risks to humanity. Without a full understanding of what makes us human—our intuitive leaps, our emotional complexities, our moral and ethical frameworks—we cannot possibly predict how a Super AI might interpret or misinterpret these very human traits. Will it share our values, or will it develop a set of objectives that are entirely alien to us? The sobering reality is that, in our current state of ignorance, we don't have the answers to these questions, making the pursuit of Super AI a profoundly risky endeavor.

The endeavor becomes even more fraught when we consider the "alignment problem," a well-known issue in AI ethics. How can we ensure that the objectives of a Super AI align with human values when we don't fully understand the depths and breadths of those values ourselves? Even with smaller-scale AI systems, we have witnessed instances where the algorithms have produced unintended and sometimes harmful outcomes due to misaligned objectives. Extrapolate this to the scale of a Super AI, and the risks become not just magnified but potentially existential.

In summary, the ethical quandary surrounding the development of Super

AI goes beyond mere technical challenges; it strikes at the very heart of our understanding of what it means to be human. Until we can answer these deeply human questions, the ethical stance must be one of extreme caution. The unknowns are not just unknown; they are ethically charged, fraught with potential peril, and demanding of our utmost intellectual and moral scrutiny.

The Singular Path to Genuine and Ethical Super Intelligence

The Natural Process of Replicating Human Intelligence

While the tantalizing notion of creating Super AI has captivated the imagination of scientists, ethicists, and futurists, it's crucial to remember that a proven method for replicating human intelligence already exists: the natural biological process of making babies. Through generations of human evolution, nature has refined the complexities of cognition, emotion, and morality into the beings we are today. As we grapple with the ethical quandaries of artificially creating an entity more intelligent than ourselves, it's worth asking: why not focus on nurturing and enhancing the intelligence that already exists in our biological offspring?

Ethical Dilemmas in Artificially Creating Human Intelligence

The ethical considerations of attempting to create artificial Super AI are daunting. This endeavor raises questions about responsibility, moral agency, and the unforeseen consequences of creating an entity that could operate outside the bounds of human understanding and control. When we don't yet fully grasp the intricacies of human intelligence, the ethical stance of trying to artificially replicate or even surpass it becomes deeply problematic.

The Singularity of Hope: A New Generation of Super-Intelligent Humans

Instead of venturing into the ethical minefield of Super AI development, why not invest in the already-existing potential for super intelligence within humanity? By embracing the concept of the "Singularity of Hope," we can focus on developing educational systems, social structures, and technologies aimed at augmenting human intelligence naturally and ethically. This would not only avoid the ethical dilemmas associated with artificial Super AI but also offer a more reliable and holistic path to genuinely intelligent entities that share our human values and ethical considerations.

In the grand narrative of intelligence, perhaps the most ethical and promising route is not in trying to create new forms of life but in enhancing existing ones. By making babies, not Super AI, we adhere to a path that respects the complexities of human intelligence and the ethical considerations that come with it.

Rethinking Super AI: Merging Mind and Machine

As we navigate through the maze of ethical, societal, and existential questions surrounding Super AI, it's important to challenge our preconceptions. The traditional vision of Super AI often conjures up images of machines operating independently of human guidance, surpassing us in intelligence and perhaps even making decisions that could jeopardize our existence. However, this perspective fails to consider a more harmonious and perhaps attainable vision: the real Super AI as the convergence of human intelligence with machine capabilities.

Drawing upon the ideas laid out in earlier chapters, particularly the limitations of machine intelligence and the untapped potential of the human mind, we propose a radical rethinking of what Super AI could be. The vision here is not of machines replacing humans but of humans and machines augmenting each other's capabilities. This is not a new concept; after all, the history of technology is filled with examples of tools that extend human capabilities, from the wheel to the smartphone. However, this symbiosis takes on a new dimension in the age of advanced machine learning and AI. We have the opportunity to create not just external tools but integrated systems—extensions of our own cognition and decision-making processes.

The real Super AI could be this synergy, where the machine learns from the human, and the human benefits from the machine's computational power, data analysis, and pattern recognition. In this collaborative model, the ethical dilemmas associated with creating an entity potentially more intelligent than humans are mitigated because the resulting super intelligence is human at its core. This is the true Super AI, where the singularity is not a point of divergence between human and machine, but a point of convergence, enhancing the best aspects of both.

Historical Perspective on Human-Machine Collaboration

Tracing Back to the Earliest Tools and Technologies
Long before the dawn of computers and artificial intelligence, humans have been in a symbiotic relationship with tools and technologies. The first stone axes extended our physical abilities, allowing us to shape the environment to our needs. The invention of the wheel, the plow, and later, complex machinery all served as extensions of human capabilities—each innovation amplifying what we could achieve.

How Each Technological Advancement Has Been an Extension, Not a Replacement
In the narrative of technology, there's a recurring theme: each new tool serves to augment human abilities, not replace them. The telescope didn't replace the astronomer; it extended their vision to the far reaches of the universe. The calculator didn't replace the mathematician; it made complex calculations more manageable and accurate. In the same spirit, the real Super AI will not replace human intelligence but extend it, allowing us to reach heights unimaginable alone.

The Limitations of Pure Super AI: Exploring the Challenges and

Unpredictabilities of Pure AGI
While the allure of creating a machine with general intelligence similar to humans is tempting, we must confront the stark limitations. Purely machine-based AGI, devoid of human guidance, is fraught with unpredictability. As discussed in previous chapters, our lack of complete understanding of human intelligence makes it implausible to create a machine that could safely surpass us.

The Nuances, Emotions, and Ethics Machines Can't Replicate
When we consider the complexities of human intelligence—our capacity for empathy, our intuitive grasp of ethical nuance, our ability to create art and music that moves the soul—it becomes evident that these are not simply data points that can be programmed into an algorithm. They are the manifestations of a complex interplay of biology, experience, and something ineffable that makes us human. While machines can mimic patterns, they can't understand the weight of an ethical dilemma, the emotional depth of a piece of art, or the complex dynamics of human relationships.
A Call to Our Humanity
In the quest for Super AI, it's crucial to remember our unique strengths as humans. We have the potential to guide AI development in a direction that amplifies the best aspects of humanity, creating a future where machines are not our replacements but our partners. The Singularity of Hope is not in a machine that surpasses us but in the co-evolution of humans and machines, each pushing the other towards greater heights. This is the vision we should strive for—a future where the true Super AI is a testament to human innovation, ethics, and spirit.

The Power of Augmentation

Diving Deep into the Concept of Human Intelligence Augmented by AI
The Singularity of Hope resides not in machines that outwit us but in systems that amplify our innate abilities. Augmentation—bolstering human intelligence with machine capabilities—represents the most

ethical and promising path forward. This concept is deeply rooted in our shared humanity, valuing our unique cognitive and emotional capabilities while using technology to extend their reach.

Real-World Examples of Augmentation Leading to Unprecedented Achievements

Consider medical diagnostics, where AI algorithms can rapidly analyze medical images but lack the clinical context that a human doctor brings to the table. When both work in tandem, the accuracy of diagnoses improves significantly. Or take the realm of scientific research. While AI can sift through vast datasets to find patterns, the human researcher poses the questions and interprets the results meaningfully, driving innovation to new heights. These are not isolated examples; they're early indicators of a future where human-machine collaboration achieves what neither could accomplish alone.

Mental Health and Augmentation: The Positive Implications of Feeling Empowered and in Control

Imagine a world where the daily frustrations of limitations are replaced by a sense of empowerment. Properly implemented augmentation can remove the tedious aspects of our lives, freeing us to spend more time in meaningful pursuits, enhancing our mental well-being.

Addressing Potential Anxieties and Concerns About Over-Reliance on Technology

While the benefits are significant, it's crucial to address human concerns about becoming too reliant on machines. The balance lies in using AI as a tool for augmentation, not a crutch that makes us forget our inherent capabilities. Keeping the human in the loop is not just good design; it's a safeguard for our collective mental health.

The Ethical Implications of Augmentation: Balancing Human and Machine Input in Critical Decisions

The ethical landscape of augmentation is intricate but navigable. In critical sectors like healthcare, jurisprudence, and creative arts, the machine's role should be to assist, not dictate. For instance, in judicial decisions, an AI could provide data on similar past cases, but the final judgment must lie with the human, who can weigh ethical and emotional factors that a machine can't comprehend.

Ensuring the Human Touch Remains Central

At the heart of ethical augmentation is the irreplaceable value of the human touch—our ability to understand context, empathize, and make decisions that respect the dignity and worth of other human beings. As we advance into an age increasingly shaped by AI, let's make it an age of augmentation, where technology enhances human capabilities while the essence of humanity remains at the core.

This is not merely a technological challenge but a deeply human one. It's an opportunity to reaffirm what makes us human—our ethics, our empathy, our creativity—and to let these qualities guide us into the future.

Preparing the Next Generation to Seamlessly Integrate with AI Tools

As we journey toward the Singularity of Hope, it becomes increasingly important to prepare the next generation for a future where human intelligence and machine capabilities are seamlessly integrated. This doesn't mean teaching kids how to code as much as it means imparting a new set of literacies. In an era where information is abundant but wisdom is scarce, the education system must evolve to equip young minds with the skills to use AI as a natural extension of their own capabilities.

Emphasizing Skills Like Critical Thinking, Empathy, and Interdisciplinary Collaboration

In a future governed by the ethics of augmentation, the most valuable skills will be inherently human. Critical thinking will empower individuals to use AI as a tool for problem-solving, not as a crutch that dulls their own analytical abilities. Empathy, an attribute no machine can replicate, will enable them to use AI responsibly, creating technologies that understand and respect human values.

Interdisciplinary collaboration is another skill of paramount importance. The challenges and opportunities of the augmented era are too complex for any single discipline to solve. Engineers will need to understand the ethical implications of the technologies they create, just as ethicists will need a working knowledge of what these technologies can and cannot do.

By placing these inherently human skills at the core of education, we're not just preparing young people to navigate an AI-augmented world; we're setting the stage for them to lead it. The future does not belong to machines or algorithms; it belongs to augmented humans who can harness machine capabilities while elevating human virtues.

This educational blueprint is more than a curriculum; it's a call to our shared humanity. It acknowledges that while machines can process information, only humans can impart wisdom. As we prepare for a future alongside increasingly capable machines, let's also prepare to bring our full, irreplaceable humanity into that future.

The Importance of Public Discourse and Awareness

As we stand on the precipice of an augmented era, the role of society in shaping our collective future cannot be overstated. The dialogues we have today will set the ethical and cultural norms of tomorrow. It is essential that these conversations are inclusive, reflecting the diverse array of human experiences and perspectives. Augmented intelligence offers a palette of opportunities, but it is through public discourse that we decide the colors to pick and the picture we want to paint.

In a world where technology evolves at a breakneck speed, adaptability is not just a useful skill; it's a survival skill. Our educational systems, our workplaces, and even our social structures must encourage a culture of continuous learning. The ability to adapt and evolve, to pick up new skills and abandon outdated practices, is what will allow us to stay ahead of the curve, not just catch up to it. Remember, in the augmented future, it is not man versus machine but man and machine versus problem.

The Myth of Replacement

Chris: "So, Taylor, what do you think about the notion that we're on the brink of mastering Super AI?"

Taylor (smirking slightly): "Mastering Super AI? Chris, that's like a child claiming they'll build a spaceship. Sure, we've made strides in AI, but understanding the full spectrum of human intelligence? We're toddlers in a cosmic playground."

Chris: "I understand the skepticism. It's true that our grasp of human intellect is still primitive. But isn't that the beauty of it? The endless pursuit, the journey of discovery?"

Taylor: "A beautiful journey, no doubt, but fraught with illusions. We're like ancient mapmakers charting unknown seas, claiming dragons where there's just fog. Our understanding of AI is still in its nautical dawn."

Chris: "But every dawn leads to daylight. No, we don't fully grasp human intelligence, which makes creating a Super AI complex. But isn't that a call for cautious exploration rather than retreat?"

Taylor: "Cautious exploration, absolutely. But let's not mistake the map for the territory. We're charting paths in AI without knowing the terrain. It's not just about humility in our approach; it's about recognizing our blind spots."

Chris: "Agreed. Humility and caution are essential. We're not gods crafting a new sentient being; we're explorers understanding a new form of intelligence. It's about respecting the unknown, not claiming dominion over it."

Taylor: "Exactly. And while we talk about augmenting human capabilities, let's not forget our own limitations. AI, for all its potential, is still a tool crafted by flawed hands."

Chris: "Flawed, yes, but capable of learning and adapting. Our journey with AI is a reflection of our own evolution– stumbling, learning, evolving. The path to Super AI isn't about achieving omnipotence; it's about enhancing our collective wisdom."

Taylor: "As long as we remember that wisdom comes with knowing what we don't know. That's the crux. We're not building a Super AI yet; we're learning how to crawl before we can walk the path of true intelligence."

Chris: "A journey of a thousand miles begins with a single step, right? And each step in AI development is a step towards understanding ourselves and the intelligence we aspire to create."

The narrative of AI as an existential threat has been a staple of popular culture for decades. Yet, if this book has argued anything, it's that machines are not here to replace us but to augment us. The real existential threat is not artificial intelligence but artificial stupidity, where we misuse the tools at our disposal.

The essence of augmentation is enhancement, not replacement. Machines can take care of data crunching, pattern recognition, and even complex simulations, freeing humans to do what they do best: think creatively, strategize, empathize, and make nuanced judgments. These are areas where AI falls short, and humans excel, the areas that make us inherently valuable.

As we continue integrating AI into our lives, let us not forget that the ultimate goal is not to build machines that think like humans but to empower humans to think better, be more creative, and solve problems more efficiently. In this endeavor, the machine is not our rival but our ally.

In summary, the future is not about the survival of the fittest but the survival of the most harmonious. Let's embrace our tools, integrate them wisely into our lives, and march towards an augmented future led not by the algorithms but by our enduring and irreplaceable humanity.

The Vision of a Harmonized Future

As we close this chapter, it is time to reflect on the transformative journey we've embarked upon—a journey where mind and machine are not adversaries but allies, and where the road to Super AI is shared. The true Super AI, as posited throughout this book, is not a standalone entity eclipsing human intelligence. Rather, it is the natural evolution of human intelligence, augmented by the computational prowess of machines.

We've explored the limitations of pursuing a purely machine-based Super AI, a quest that not only presents ethical and existential dilemmas but also sells short the unique attributes of human cognition. Machines may excel at computation, but they lack the emotional depth, ethical considerations, and creative spontaneity that define human intelligence.

Instead of being entrapped by the mirage of a machine-dominated future, let's focus on the real oasis: a future of augmented intelligence, where humans and machines collaborate to solve the world's most pressing challenges—from climate change and poverty to healthcare and education. Imagine a future where augmented doctors make more accurate diagnoses, where augmented educators tailor learning to individual student needs, and where augmented policymakers make more informed and ethical decisions.

The synergy between human intelligence and machine capabilities promises to usher us into a new era of unprecedented progress and well-being. It will be an era characterized not by surrendering our humanity to machines but by elevating our humanity through them.

But for this future to materialize, we must exercise both caution and enthusiasm. Caution to ensure that our ethical and human-centric considerations guide every step of our technological advancement and enthusiasm to explore the infinite possibilities that such a harmonious partnership can bring forth.

Our journey towards this augmented future is only just beginning, and the path may be fraught with challenges and ethical quandaries. Yet, it is a journey worth undertaking—a journey that calls upon our highest faculties of reason, ethics, and imagination. For in the harmonious fusion of man and machine, we glimpse the extraordinary potential of the true Super AI: a future not overshadowed by machines but enlightened by our irreplaceable and ever-augmenting humanity.

In the age of disruptive technologies and rapid digital transformation, the "Singularity of Hope" stands as an anchor for humanity. It is a vision where the convergence of mind and machine doesn't dilute our human essence but amplifies it, creating a harmonious future that is greater than the sum of its parts. The mutual benefits are compelling: machines gain the richness of human context, while humans gain the computational power to navigate an increasingly complex world.

Now, let's elevate these notions to the context of the new Super AI, a harmonious blend of human and machine intelligence. In this envisioned future, the AI taxpayer and investor are not just algorithmic entities making calculations based on data; they are integrated systems that make decisions with ethical and social implications, guided by human values and intuition.

This harmonized vision extends beyond just economic or social systems. Imagine AI-enhanced healthcare where machines provide data-driven insights but where human professionals make the final call, always with a touch of empathy and ethical consideration. Picture a future of augmented creativity, where artists and AI collaborate to produce art that resonates both intellectually and emotionally.

The true Super AI in the "Singularity of Hope" is not a monolithic entity but a collective intelligence, a blend of the best of what both humans and machines have to offer. It is an entity of unparalleled problem-solving capabilities, yet grounded in the ethical and emotional context that only humans can provide.

Such a future is not just a fanciful vision but a call to action. It invites us to redefine what is possible, to look beyond the binaries of human vs machine, and to see the extraordinary potential that lies in our combined strengths. It's a call for unity in a world of apparent divisions, a call for hope in times of uncertainty, and most importantly, a call to safeguard and celebrate our shared humanity as we journey into the unknown.

In this harmonious future, we don't just adapt to technological change; we thrive in it. And that is the most beautiful vision of all, one where the true Super AI is a manifestation of our highest ideals and where the Singularity brings not an end but a hopeful new beginning.

References:

1. Kurzweil, Ray. The Singularity is Near: When Humans Transcend Biology. Viking, 2005.

2. Russell, Stuart. Human Compatible: Artificial Intelligence and the Problem of Control. Viking, 2019.
3. Bostrom, Nick. Superintelligence: Paths, Dangers, Strategies. Oxford University Press, 2014.
4. Li, Fei-Fei. "AI and Our Future." Talk at TED Conference, 2017.
5. Vance, Ashlee. Elon Musk: Tesla, SpaceX, and the Quest for a Fantastic Future. Ecco, 2015.
6. Hassabis, Demis. Interviews and talks available on DeepMind's official YouTube channel.
7. Buolamwini, Joy. "Gender Shades: Intersectional Accuracy Disparities in Commercial Gender Classification." Proceedings of Machine Learning Research, 2018.
8. Tegmark, Max. Life 3.0: Being Human in the Age of Artificial Intelligence. Knopf, 2017.
9. Kasparov, Garry. Deep Thinking: Where Machine Intelligence Ends and Human Creativity Begins. PublicAffairs, 2017.
10. Ng, Andrew. "AI is the New Electricity." Speech at Stanford Graduate School of Business, 2017.
11. LeCun, Yann. "Our Intelligence is What Makes Us Human, and AI is an Extension of that Quality." Keynote at Neural Information Processing Systems Conference, 2017.
12. Berners-Lee, Tim. "The Next Web." TED Talk, 2009.
13. Wojcicki, Susan. "Concerns About AI." Interview at SXSW Conference, 2018.
14. Horvitz, Eric. "We are Stewards of This Planet." Interview with GeekWire, 2018.
15. Dignum, Virginia. Responsible Artificial Intelligence: How to Develop and Use AI in a Responsible Way. Springer, 2019.
16. Nadella, Satya. "AI and Ethics." Speech at Microsoft AI Event, 2018.
17. Brooks, Rodney. "The State of AI." Interview with ZDNet, 2019.
18. Marcus, Gary. Rebooting AI: Building Artificial Intelligence We Can Trust. Pantheon, 2019.

19. Bengio, Yoshua. "The Future of AI." Interview with Nature, 2019.

Conclusion: A Call to Humanity in the Age of Machines

As we stand on the precipice of a new era, a world flooded with algorithms and machines, we find ourselves not at the end of a journey but at the dawn of a movement. This is not just a conclusion; it's a call to action for humanity. We now wish to distill the essence of our odyssey through "The Singularity of Hope," transforming insights and visions into a compelling rally to arms.

Recap of the Journey

As we draw the curtains on our exploration in "The Singularity of Hope," let us pause and reflect upon the narrative we've seen play out. Our odyssey began with exploring the rich tapestry of human intelligence, juxtaposing it against the fledgling capabilities and bounds of AI. We journeyed through the realms of AI in the workplace and marveled at the promise of HAIA, navigating the intricate dance between technology and our human essence. Discussions on the AI singularity peeled back layers of future possibilities while underscoring the indispensable role of human guidance in the evolution of AI.

The Imperative of Human Guidance in AI

A resounding truth emerged in our quest: the paramount necessity for human intellect to shepherd AI. The AI of our times, in its nascent state, falls short of mirroring the full spectrum of human cognition. It behooves us, therefore, to wield AI with wisdom and foresight, ensuring it acts as a complement to, not a replacement for, our uniquely human faculties.

Actionable Steps for Ethical AI

To steward the ship of ethical AI, we must embark on a path that includes:

- Upholding transparency in AI's algorithms and decisions.
- Cultivating interdisciplinary synergy to unravel and address the ethical conundrums posed by AI.
- Implementing robust data governance, guarding privacy, and ensuring fairness.
- Committing to the continuous ethical consideration of those who mold AI.

Role of Different Stakeholders

This mission calls for a unified front across the spectrum of our society:

- Technologists, as architects of AI, must pledge to its responsible creation.
- Policymakers are tasked with sculpting regulations that foster ethical AI utilization.
- Educators bear the torch of integrating AI ethics into the bedrock of learning.

- Citizens, as the bedrock of society, should stay informed and actively participate in dialogues surrounding AI's role in our shared destiny.

Establishing Ethical Frameworks

To navigate AI's voyage responsibly, we must erect ethical frameworks and governance models. This endeavor encompasses crafting policies that ponder the long-term ramifications of AI, assuring that its boons are shared equitably and its potential perils are meticulously managed.

Collaborative Efforts for a Better Future

The future tapestry of AI will be woven from the threads of collaborative efforts. It beckons us to foster alliances across industries, academic sanctuaries, and government halls. It's about fostering a symposium of diverse perspectives, converging to chart the ethical course of AI.

The Divine Spark Within Us

In this age of AI, it's crucial to remember that though machines may roam our streets, they cannot replicate the divine spark within us. They lack the capacity to feel, to love, to create art from a place of deep emotion. It's in this realization that our true power lies. As we navigate this new world, it's not about being overshadowed by AI; it's about harnessing our unique human spirit to shine even brighter.

Walking Hand-in-Hand with Machines

AI didn't take our jobs; it took our routine and gave us back the creativity—the canvas on which human innovation thrives. We are called not to battle against these machines but to dance with them. AI should not replace us but rather complement and amplify our capabilities. It handles the mundane while we, the blessed beings, infuse creativity, emotion, and ethical grace into our endeavors. In the Singularity of Hope, prosperity was not an outcome but a journey—one taken by all of humankind, hand in hand with their AI robots,

Work, Redefined and Reimagined

As jobs evolve, let's embrace this metamorphosis. Let AI take on computational tasks while we focus on dreaming, feeling, and connecting. The value we bring lies in our hearts and souls, not just in the tasks we perform. As the machine learned, so did we. But while it accumulated data, we gathered wisdom. The difference was stark yet symbiotic.

Educate for the Soul, Not Just the Role

Our education systems must evolve to nurture deep thinking, ethical actions, and profound emotions. In a world intertwined with machines, we must prepare our young not just for jobs but for meaningful, soulful engagements in a post-AI world.

Community– Our Beacon in the Digital Storm

In the vast digital landscape, our human connections remain our strongest anchor. Our communities provide a sense of purpose and belonging, helping us navigate the AI era with resilience and hope. The true singularity was not a point in time but a point of perspective—a shared vision for a prosperous world.

Healing the Wounds of the Soul

We must address the mental battles intensified by AI's rise – the shadows of doubt, anxiety, and redundancy. By anchoring ourselves in purpose and community, we can fortify our minds and find healing in meaningful engagement.

A Sanctuary of Hope in a Chaotic World

In this turbulent era, we create sanctuaries of hope by embracing our humanity and our role alongside AI. These sanctuaries act as buffers against mental tumult, every purposeful action and heartfelt connection fortifying our well-being.

As we conclude, we present a call for all aspects of society to stand united–technologists, policymakers, educators, and citizens–ready to shape a future where AI and human intelligence synergize. This conclusion is our blueprint for a world where technology amplifies humanity. We close this chapter only to open a new one in human history, where we embrace AI's potential with wisdom, foresight, and a commitment to the betterment of all. The singularity is not an end but a doorway—leading to corridors of innovation, prosperity, and unforeseen kindness.

The rise of AI is not our end but our new beginning. A chance to redefine and rise with purpose, passion, and soulful engagements. This age will be known as the age in which AI didn't replace the human touch; it will be known as the age when AI amplified it, turning individual acts of kindness into global movements. Let's march forward with hope, purpose, and unwavering faith in our shared destiny!

Appendix A: The Ethical Matrix: Navigating Rights and Responsibilities

Introduction: The Evolution of AI Ethics

In the fascinating trajectory of technological evolution, perhaps no development has been as consequential and as awe-inspiring as Artificial Intelligence (AI). Yet, as these intelligent algorithms continue to weave their way into the very fabric of our society, the ethical ramifications of their integration become ever more prominent. The field of AI ethics is not just an academic exercise—it's a vital endeavor that will influence the course of our collective future.

AI ethics has emerged as a critical area of concern for multiple reasons. Foremost, AI systems, by their very nature, make decisions. These decisions, whether in healthcare diagnostics, financial lending, or even entertainment recommendations, have tangible impacts on individuals and communities. As AI's decision-making capacities grow, so do the complexities of ensuring these decisions are made fairly, transparently, and with respect to human dignity.

Moreover, AI doesn't exist in a vacuum. It is shaped by human inputs, influenced by cultural biases, and governed by business interests. Hence, ensuring ethical AI isn't just about tweaking algorithms—it's about understanding the broader socio-cultural milieu in which AI operates. It's about recognizing that AI is just as much a product of human society as it is a force shaping it.

Emerging trends in AI ethics reflect a growing awareness of these intertwined complexities. From renewed approaches to data privacy and consent, to ambitious strides in algorithmic accountability, there's a palpable momentum towards forging a more ethical AI landscape. Collaboration, once seen as a mere facilitator, is now viewed as a cornerstone of ethical AI, bringing together diverse voices and disciplines. Governance, too, has come to the fore, with localized strategies being balanced against the quest for global standards.

As we embark on this exploration, it's vital to remember that AI ethics is not a static field. It's a continuous, dynamic dialogue—a collective journey that requires the engagement of technologists, policymakers, ethicists, and indeed, every one of us.

Algorithmic Accountability

As AI systems continue to influence decisions that shape individual lives and societal structures, their mysterious and often opaque nature has led to mounting concerns. This "black box" phenomenon, where algorithms process data and produce outcomes without transparent explanations, poses profound ethical challenges. At the heart of these concerns is a simple yet powerful demand: accountability.

Beyond the Black Box: Making AI Understandable

Our reliance on AI is accompanied by an inherent trust that the systems will operate in our best interests. However, when the workings of these systems are shrouded in complexity, that trust can quickly erode. To bridge this trust deficit, the AI community is increasingly focusing on interpretability and transparency.

Interpretable AI refers to models that are designed to be inherently understandable. While this may sometimes come at the cost of model accuracy, the trade-off ensures that every decision made can be explained in terms humans can grasp. Transparent AI, on the other hand, emphasizes understanding the decision-making process. Techniques such as Local Interpretable Model-Agnostic Explanations (LIME) or SHAP (SHapley Additive exPlanations) values are becoming popular in demystifying how AI systems reach particular conclusions.

Holding Algorithms Accountable

Accountability in AI is not merely a technical challenge,it's also very much an ethical and legal one. The question "Who is responsible when AI goes wrong?" has multi-faceted answers.

Developers and Companies: AI developers have an onus to ensure their models are not perpetuating harmful biases or making erroneous decisions. Regular audits, ethical evaluations, and continuous training can help mitigate potential pitfalls.

Regulators: Governments and international bodies can play a pivotal role by setting standards and guidelines for AI systems, ensuring that there are tangible repercussions for breaches of trust or harm.

End-users: There's also an emerging perspective that users should be educated about the AI systems they interact with, allowing them to be vigilant watchdogs, holding systems accountable by demanding transparency and rectification when anomalies occur.

Challenges in Ensuring Accountability

Evolving Nature of AI: Unlike traditional software, AI systems can learn and evolve. This dynamic nature makes pinpointing accountability challenging, as the system's decisions may change over time.

Scale: The vast scale at which some AI models operate, analyzing millions of data points, can make it challenging to track and understand every decision.

Trade-offs: As mentioned, there's often a trade-off between interpretability and performance. Highly accurate models might be more complex and less interpretable.

As AI systems become increasingly ingrained in the socio-economic fabric, ensuring their accountability is not just desirable—it's imperative. The path to accountable AI is complex, demanding technical innovations, ethical deliberations, and regulatory oversight. But at its core, it's about reinforcing a foundational principle: AI, no matter how advanced, should serve humanity responsibly, ethically, and transparently.

Data Privacy and Consent

In our digital age, data is often likened to oil—central to powering technological innovations, yet fraught with controversies regarding its collection, storage, and usage. Artificial Intelligence, in particular, thrives on vast amounts of data. Yet, as AI's appetite for data grows, so does the urgency to address pivotal questions of privacy and consent.

The New Face of Data Privacy

Gone are the days when data privacy was a mere afterthought, discussed in the appendices of tech policy documents. Today, it occupies center stage. Regulatory frameworks like the European Union's General Data Protection Regulation (GDPR) have set groundbreaking standards, emphasizing individual rights over personal data. But as AI continues to evolve, so does the need to redefine and expand upon these standards.

AI systems often operate in realms beyond traditional data categories. For instance, while one may willingly share their name or address, how do we account for data like behavioral patterns, facial expressions, or even emotional states—intangible yet deeply personal attributes that AI can infer or predict? The challenges lie not just in protecting data but in comprehensively understanding what constitutes 'data' in the AI context.

Informed Consent in the Digital Age

Historically, consent in the digital realm has been a check-box exercise, often buried within lengthy terms and conditions. But as AI systems grow more complex, the implications of granting consent become profound. It's no longer enough to just inform users that their data will be used; it's imperative to communicate how AI will use, interpret, or even transform this data.

Transparent AI demands a reimagining of consent. Users must be provided with clear, understandable insights into AI's decision-making processes. They should know if their data is feeding a machine learning model, influencing an algorithm, or even being shared across platforms. This isn't just a legal necessity; it's an ethical imperative.

Moreover, the static, one-time nature of consent is being challenged. In a world where AI systems learn and evolve, periodic re-consenting and active user engagement become crucial. It's about fostering a dialogue, not just securing a one-off permission.

In our interconnected world, data flows are intricate, vast, and incessantly dynamic. As AI stands at the crossroads of this data-driven ecosystem, upholding privacy and securing informed consent aren't just challenges—they're prerequisites for an ethical AI future. It's about recognizing the inherent human dignity behind every byte of data and ensuring that as AI systems learn more about us, they do so with respect, transparency, and integrity.

Collaboration in AI Development

The journey of AI ethics and accountability cannot be traversed by isolated entities. It requires a confluence of perspectives, expertise, and shared objectives. Collaboration emerges as a linchpin, creating a holistic ecosystem where AI evolves responsively and responsibly.

The Multidisciplinary Nature of AI Ethics

AI, at its essence, is not merely a technological endeavor. It intersects with sociology, philosophy, law, and many other disciplines. Ensuring AI is ethical and just requires pooling knowledge from these diverse areas. For instance:

Philosophers can help navigate moral dilemmas that AI presents.
Sociologists can provide insights into how AI affects societal structures.
Legal experts can guide on regulatory frameworks.
Collaborative projects, like OpenAI's Charter which emphasizes cooperation over competition, can showcase the power of combined expertise in shaping AI's future.

Public and Private Sector Partnerships

Marrying the innovative drive of the private sector to the public sector's regulatory oversight can accelerate the development of ethical AI. Such collaborations can:

- Streamline regulations, ensuring they're both innovative-friendly and protective of public interest.
- Facilitate public-funded research, democratizing AI's benefits.
- Create shared standards and best practices across the industry.

Examples like the Partnership on AI, which brings together tech giants and non-profits, exemplify the promise of these cross-sector collaborations.

Governance in AI: From Local to Global

In a world interconnected by technology, the governance of AI remains a complex tapestry of localized needs balanced against universal principles.

Localized AI Governance Strategies

Different cultures, societies, and economies will have unique relationships with AI. Tailoring AI ethics and regulations to these specific contexts is crucial. For instance:

In countries with robust data infrastructure, data privacy laws might focus on sophisticated cyber threats.
In others, the primary concern might be basic data access or literacy.

The Quest for Global AI Governance

While localization is key, the borderless nature of digital technology makes global governance indispensable. A fragmented AI landscape, where each region operates in silos, can hinder AI's potential. Efforts are underway to forge universal standards.International bodies like the UN and WHO are initiating conversations about global AI norms.
Consortia of tech companies are coming together to draft shared AI principles.
Despite its challenges, global governance is not just about uniformity; it's about ensuring that as AI impacts the world, no region is left behind or unfairly disadvantaged.

Collaboration and governance, both local and global, are shaping the contours of AI's evolution. The synergy of diverse minds and collective aspirations ensures that AI, as it strides into the future, does so hand-in-hand with humanity's shared ethical values.

Preparing for the Unforeseen: Ethical Safeguards

As AI continues its meteoric rise, its complexities become increasingly intertwined with the nuances of the human experience. While we aspire to predict and control every facet of this revolutionary technology, the reality is that we must also be prepared for the unforeseen. The very nature of AI's learning capabilities implies unpredictability. Thus, it becomes paramount to embed ethical safeguards, ensuring AI remains a tool for progress, not a harbinger of unintended consequences.

Anticipatory Ethics: The Forethought Approach

Rather than reacting post facto to AI's ethical dilemmas, an anticipatory approach seeks to forecast potential challenges and preemptively address them. This involves:

Scenario Planning: Regularly brainstorming and simulating possible challenges AI might pose in diverse sectors, from healthcare to finance.
Ethical Stress Testing: Subjecting AI systems to a variety of ethical dilemmas to observe their responses and refine them accordingly.

Moral Guardrails: Embedding Ethical Constraints

One practical way of ensuring ethical behavior in AI is by setting clear moral guardrails within which the system operates. This can include:

Value Alignment: Ensuring the AI's objectives are closely aligned with human values and societal norms.
Boundaries: Setting clear operational parameters beyond which the AI doesn't venture, such as not allowing it to make decisions in areas where human judgment is crucial.
Feedback Loops: The Continuous Learning Process

AI's learning shouldn't be isolated. Establishing feedback mechanisms allows AI to continually refine its ethical compass. This involves:

Human Oversight: Incorporating human intervention, especially in the

early stages, to guide AI's decision-making.
Community Reporting: Allowing end-users to report perceived ethical breaches or biases in AI's functioning, ensuring the system is accountable to the very populace it serves.
Transparency and Open Source: The Power of Collective Vigilance

By making AI algorithms transparent and, where appropriate, open source, we invite collective scrutiny. This communal oversight can:

Spot Flaws: The broader the reviewing community, the higher the chances of catching biases or ethical flaws.
Democratize AI Ethics: An open approach ensures diverse voices contribute to shaping AI's ethical framework.

The journey of AI is as much about pioneering technological advancements as it is about navigating the uncharted ethical waters. By putting in place robust ethical safeguards, we can ensure that even in the face of the unforeseen, AI remains steadfastly anchored to our shared moral compass, championing progress without compromising principles.

The Collective Journey Towards Ethical AI

As our odyssey into the age of artificial intelligence deepens, it becomes clear that the boundaries between machine capabilities and human intelligence aren't starkly defined. Instead, they merge, intertwine, and co-evolve in a dance of synergy. This dance, as we've explored throughout this book, isn't merely about efficiency or productivity; it's a dance of ethics, values, and shared purpose.

The singularity of hope, a vision we've consistently revisited, isn't merely a symbol of technological advancement but a beacon of ethical maturity. It serves as a reminder that as we augment our capabilities with AI, we must also augment our moral compass, ensuring it remains attuned to the ever-evolving challenges and opportunities this merger brings.

Fear, as we've discussed, is the "mind-killer." And while it's natural to harbor apprehensions about the unknown terrains of AI, it's crucial to remember that the real enemy isn't the technology itself, but ignorance and complacency. AI, in its essence, is like a mirror, reflecting back our own values, biases, hopes, and fears. If there are shadows in this reflection, they aren't born out of the machine but from the depths of our collective consciousness.

Therefore, the call to action isn't just for technologists, policymakers, or business leaders. It's a collective clarion for humanity. Every individual has a stake in this future, and every voice matters in this dialogue. Just as the augmentation approach emphasizes the harmonious collaboration between man and machine, the journey towards ethical AI must be a collective endeavor, involving diverse perspectives, interdisciplinary insights, and a shared commitment to upholding human dignity, rights, and well-being.

Our work is far from done. Each day brings new challenges, dilemmas, and opportunities. But as we've seen, with a mindset anchored in augmentation, a commitment to continuous learning, and a heart attuned to ethical considerations, we can navigate this journey with wisdom, empathy, and hope.

In the age of AI, our greatest strength isn't just our augmented intelligence but our augmented humanity. As we stand at this pivotal juncture, let us move forward with a renewed pledge to harness AI's potential not just for technological advancement but for the betterment of all, ensuring that the singularity we approach is indeed one of hope, unity, and shared ethical prosperity.

The Need for Responsible AI Design

In our exploration of the intricate dance between the human mind and machine intelligence, we've consistently returned to a central tenet: the augmentation of human capabilities. But as AI systems become more sophisticated, weaving themselves into the very fabric of our daily lives, the importance of designing these systems responsibly becomes paramount. This isn't just about efficiency or integration; it's about crafting AI that resonates with our shared human values, ethical principles, and the collective vision we've termed the "singularity of hope."

AI, as we've often reiterated, is not an alien entity but a reflection of our collective consciousness. It learns from our data, mirrors our biases, and amplifies our intentions. If this reflection is to remain true to our highest aspirations and not our deepest fears, then the design of AI systems must be approached with utmost responsibility.

One of the pressing challenges we face today in the realm of AI is the issue of transparency. As AI algorithms operate in increasingly complex domains, their decision-making processes can become convoluted, leading to what many term the "black box" phenomenon. When we don't understand how an AI system reaches its conclusions, it becomes challenging to trust it, let alone align it with our ethical values.

Moreover, the augmentation approach, which we've championed throughout this book, underscores the importance of collaboration between humans and AI. For this collaboration to be effective, humans need to understand, at least to a reasonable extent, how AI systems operate. This understanding fosters trust, enables effective human-AI teamwork, and ensures that the AI acts in ways that are consistent with our shared values.

Additionally, the rapid advancements in AI have outpaced regulatory and ethical frameworks. While the machine learns at an exponential rate, our societal structures, rooted in centuries-old paradigms, often struggle to keep up. This lag creates gaps—gaps in accountability, transparency, and responsibility.

To bridge these gaps, there's a pressing need for value-centric AI design. Such a design would prioritize ethical considerations from the outset, ensuring that AI systems are built with an inherent respect for human rights, dignity, and values. Furthermore, explainable AI, which demystifies the decision-making processes of complex algorithms, can play a pivotal role in fostering trust and accountability.

In conclusion, as we continue our journey towards the singularity of hope, it's crucial to remember that every step taken in AI development should be a step forward in ethical responsibility. By championing responsible AI design, we not only safeguard our present but also lay the foundation for a future where machines don't overshadow humanity but illuminate its highest potential.

Roles and Responsibilities in AI Development

In our collective journey towards a future where the confluence of human and machine intelligence leads to a "singularity of hope," it's vital to recognize the roles and responsibilities of key stakeholders in AI development. These stakeholders each play a distinct, yet interrelated role in ensuring that AI systems are designed, deployed, and utilized responsibly and ethically.

AI Developers and Engineers

AI developers and engineers stand at the frontline of this technological revolution. As architects of the AI systems that are poised to reshape our world, they bear the significant responsibility of ensuring that these systems are ethically designed and operate within a framework that emphasizes human augmentation, not replacement.

However, it's crucial to understand that the task of creating ethical AI isn't one that rests solely on developers' and engineers' technical prowess. It requires an interdisciplinary collaboration—merging the technical with the philosophical, the computational with the ethical. Developers, thus, must actively seek input from ethicists, sociologists, and other experts outside their immediate domain. This collaboration ensures that AI doesn't just function correctly but also aligns with the broader human values we've emphasized throughout our exploration.

Policymakers and Regulators

The rapid pace of AI advancements underscores the need for robust regulatory frameworks. Policymakers and regulators play a pivotal role in creating a legal environment that promotes explainable and responsible AI. Their challenge is two-fold: they must ensure that AI systems adhere to ethical and safety standards, while also allowing room for innovation and growth.

In the context of our book's themes, it's imperative for these frameworks to prioritize the augmentation approach. Regulations should be designed to ensure that AI acts as a complement to human capabilities, rather than a competitor. By doing so, we can foster a landscape where AI serves to elevate human potential, rather than diminish it.

Moreover, in this evolving landscape, policymakers need to be agile. They must be prepared to revise and adapt regulations in response to new challenges and insights, always with the overarching goal of steering AI towards the singularity of hope.

End-Users and the General Public

As AI becomes more ubiquitous, its influence extends to virtually everyone, from tech-savvy early adopters to the general public. As end-users of AI systems, individuals have a role in understanding, questioning, and even shaping how these tools are applied in their lives.

However, for the public to effectively play this role, education is key. A foundational understanding of AI's workings, its potential, and its limitations empowers individuals to make informed decisions about the AI systems they interact with. Additionally, a well-informed public can actively participate in societal dialogues about AI's ethical and societal implications, helping shape its trajectory.

In the spirit of our book, it's crucial to recognize that AI, in its essence, is a tool—a "giant spell checker" if you will. Just as we wouldn't blindly accept every recommendation from a spell checker, we shouldn't uncritically accept every output from an AI system. By fostering a sense of curiosity, critical thinking, and active engagement, we can ensure that the narrative of AI remains one of augmentation, collaboration, and shared progress towards a hopeful future.

Building a Transparent and Value-Centric AI Future

As we journey through the intricate tapestry of artificial intelligence, weaving narratives of augmentation, singularity, and hope, we stand on the precipice of a transformative era. The horizons of our AI-driven future are vast, but they are not preordained. The choices we make today, as developers, policymakers, end-users, and members of the global community, will shape the contours of this new world.

Central to our explorations in this book has been the notion of AI as a "giant spell checker"—a tool that complements, not competes with, human capabilities. Just as the spell checker aids in refining our written expressions without undermining the essence of our message, AI should augment our abilities, amplifying the best of human intelligence and creativity. But for this vision to materialize, transparency and value-centricity are paramount.

Transparency ensures that AI systems are understandable and explainable. In a landscape marked by complex algorithms and vast data networks, deciphering the "how" and "why" behind AI decisions is more than a technical imperative—it's an ethical one. Without transparency, trust erodes, and the chasm between human and machine widens. With it, we foster an environment where AI is not an enigmatic oracle but a collaborative partner.

Yet, transparency alone is not sufficient. Our AI systems must also be grounded in human values—a value-centric approach that places human welfare, dignity, and aspirations at its core. As we've navigated the concept of the "singularity of hope," we've underscored the idea that technology should serve humanity's noblest ideals. In the AI epoch, this translates to systems that respect human rights, enhance well-being, and, most importantly, recognize the irreplaceable nuances of human judgment, emotion, and ethics.

The road ahead is uncharted, filled with challenges and opportunities in equal measure. But if history has taught us anything, it's that humanity has an indomitable spirit, an ability to confront the unknown with courage and ingenuity. As we stand at the crossroads of the AI revolution, let this be our call to action: to build AI not as our successor, but as our ally. To ensure that the singularity we approach is one of hope, enlightenment, and shared prosperity. And to remember, always, that our greatest enemy is not AI, but fear and ignorance.

Together, let's craft an AI future that resonates with our highest aspirations, where the harmonious dance between human and machine charts a course towards a brighter, more equitable tomorrow.

Recognizing Biases in AI

As we advance toward a future shaped by the "Singularity of Hope," the integration of artificial intelligence (AI) into various facets of society becomes increasingly imminent. However, a crucial aspect of this hopeful vision for AI involves ensuring that these systems are equitable and fair. One significant hurdle in this regard is the issue of biases in AI, which, if not adequately addressed, could amplify existing societal inequalities.

Explanation of How AI Can Inherit and Amplify Human Biases

AI systems learn from data, often generated by human activities. These data sets can contain embedded prejudices or skewed perspectives that reflect historical or social biases. When AI systems are trained on such data, they can inherit and even amplify these biases, making decisions that perpetuate inequality or discrimination. In a future where AI plays a substantial role in decision-making, unexamined biases can have grave consequences, diluting the promise of a harmonious human-machine symbiosis.

In the "Singularity of Hope," we envision a future where AI not only augments human capabilities but also promotes social equity. Addressing biases is not just a technical challenge but a moral imperative. By creating AI systems that recognize and counteract biases, we can build a society where technology reinforces the best of human values, rather than exacerbating our flaws.

Origins of Biases in AI Systems

Data-Driven Biases

Biases in AI often originate from the training data. For instance, facial recognition technologies trained on data sets that lack diverse representation can misidentify individuals from underrepresented groups. In a world where AI augments human decision-making, such biases can have real-world implications, from skewed hiring practices to unfair law enforcement.

Algorithmic and Design Biases

Beyond the data, biases can be introduced during algorithm development or model design. For example, if an AI system designed to filter job applicants is not programmed to consider varied life experiences, it might favor candidates from more traditional backgrounds. Such unintentional oversights can lead to design flaws that only serve to further existing inequalities.

The Cautious Optimism for Equitable AI

As we move closer to realizing the "Singularity of Hope," it's crucial to continuously engage with the ethical dimensions of AI. Recognizing and addressing biases in AI is a collective responsibility that extends from developers and policymakers to end-users. By doing so, we take a significant step toward a future where AI is not just powerful but also equitable and just.

In a world rapidly advancing towards the singularity, our true enemy is not the machine but our own biases and limitations. By confronting these biases now, we pave the way for a future where AI can be a force for good, amplifying the best aspects of human intelligence and compassion.

Impacts of Biased AI

As we stand on the threshold of the "Singularity of Hope," a world where human intelligence and machine capabilities coalesce into a harmonious existence, it's essential to recognize the potential pitfalls. One such pitfall is biased AI, which can have profound socio-economic ramifications if left unchecked. In this section, we delve into the possible societal impacts of biased AI and explore a case study that highlights the urgency of this issue.

Biased AI systems have the potential to exacerbate existing social and economic disparities, creating a future that diverges from the hopeful singularity we envision. For instance, biased AI in financial systems could perpetuate cycles of poverty, while AI algorithms in healthcare could lead to unequal access to medical services.

Healthcare Disparities

AI algorithms in healthcare, if improperly designed, could prioritize treatments for groups that are overrepresented in training data, thereby neglecting underserved communities. Imagine an AI-powered diagnostic tool trained predominantly on data from a particular ethnic group. The tool may become less accurate for individuals outside that group, leading to misdiagnoses and inadequate treatments—consequences that could be a matter of life and death.

Education and Skill Development

AI's role in personalized education can either be a boon or a curse. If training data lacks representation, AI-empowered education systems might cater primarily to learning styles predominant in privileged communities. This neglects learners who might benefit from alternative pedagogical approaches, hindering social mobility and perpetuating cycles of poverty.

Financial Services and Access to Credit

Example: The Impact of Biased AI in Mortgage Lending

In a future where AI augments human decision-making in financial institutions, imagine a mortgage lending AI trained on historically biased data. This AI might be less likely to approve loans for applicants from low-income neighborhoods, perpetuating a cycle of poverty and limiting opportunities for upward mobility.

In this hypothetical scenario, the AI system, designed to streamline the lending process, inadvertently becomes a gatekeeper that enforces historical biases. Families from disadvantaged backgrounds find themselves locked out of homeownership, an essential pathway to financial security. This case underscores the importance of scrutinizing training data and algorithmic design, as even seemingly neutral technologies can reinforce systemic inequalities.

As our earlier case study highlighted, biased AI in financial services like mortgage lending could create a vicious cycle where people from marginalized communities are perpetually disadvantaged. But it doesn't stop at mortgages; it extends to credit scoring, insurance premiums, and even investment opportunities. If AI systems, acting as the new gatekeepers of financial health, carry the prejudices of their training data, they can systematically lock out large swaths of the population from financial growth opportunities.

The Imperative for Equitable Algorithms

The socio-economic ramifications of biased AI stand in stark contrast to the ideal of a "Singularity of Hope," where AI augments human potential across all social strata. As we move closer to this singularity, our ethical responsibility is not just to develop AI systems that are technically proficient, but also socially and economically equitable. By doing so, we will better ensure that the road to the singularity is one that leads to a future rich in both technological marvels and social justice.

Towards a More Equitable AI Future

As we navigate the complexities of an AI-augmented future, the "Singularity of Hope" hinges on our collective responsibility to ensure equitable outcomes. Addressing the biases in AI systems is not just a technical challenge but an ethical imperative. By doing so, we pave the way for a society where AI serves as a tool for amplifying human intelligence and empathy, rather than deepening divisions.

The road to the singularity offers us an opportunity to reset, to rebuild systems that reflect our highest ideals. But to reach that future, we must first confront and rectify the biases that have been encoded into our machines. Only then can we achieve a singularity that embodies not just technological advancement, but also social progress and human compassion.

Erosion of Trust: The Invisible Barrier to the Singularity of Hope

In a journey toward what we've termed the "Singularity of Hope," the symbiosis of human intelligence and artificial capabilities, the trust factor plays a critical role. This trust isn't just a binary equation between human and machine; it's a multi-faceted relationship that involves society at large. When biased AI systems creep into the framework, they can erode this delicate trust, creating not just skepticism but potentially irreversible setbacks.

Undermining Faith in Technology

If AI systems, designed to augment human capacities and create a harmonious future, become agents of discrimination or inequity, the disillusionment can be profound. This sense of betrayal would not only impact how end-users interact with specific applications but could result in a cascading effect, making individuals question the broader role and reliability of AI. Within the conceptual frame of our book, this could be likened to a "dark shadow" of the singularity, where instead of expanding human potential, AI becomes a barrier.

Fanning the Flames of Social Unrest

Biased AI doesn't just operate in a vacuum; it interacts with existing social and political structures, often magnifying latent tensions. Consider an AI system deployed in law enforcement that disproportionately targets certain ethnic groups. Such an algorithm isn't merely an abstract technical issue; it becomes a spark for social unrest, fueling protests and even potential retaliation against technological advancements. This creates a vicious cycle: the very technology that is supposed to facilitate progress becomes a catalyst for regress, which is antithetical to the idea of a "Singularity of Hope."

Restoring Trust as a Collective Endeavor

The erosion of trust due to biased AI systems is more than a technological problem; it's a societal dilemma that demands multi-stakeholder interventions. It's not sufficient to merely rectify the algorithms; we need to rebuild the broken trust. In the road to the singularity, this involves transparent communication, ethical governance, and, above all, an unwavering commitment to social equity. Only by restoring trust can we hope to achieve a true Singularity of Hope, where AI serves as a force multiplier for human potential, rather than a divider.

Strategies for Fairness and Equity: Crafting the Blueprint for an Enlightened Singularity

In the pursuit of a Singularity of Hope, fairness and equity stand as non-negotiable cornerstones. The augmentation of human intelligence through AI should not be a privilege but a universal empowerment. How can this be achieved? The answer, or at least a significant part of it, lies in how AI systems are trained and developed.

Inclusive Data Collection: The Seed of Equitable AI

The notion of "Garbage In, Garbage Out" is well-known in computing. In the context of AI and mental health, it takes on a crucial meaning: biased or unrepresentative data can lead to AI systems that are not just unfair but also potentially harmful. If we are to trust AI systems with our mental well-being, the data that train these systems must be beyond reproach.

Diversity as a Mandate, Not an Option

Inclusive data collection means going beyond the low-hanging fruits. It requires proactive steps to ensure that the data used to train AI systems are representative of the diverse populace they serve. This involves not only ethnic and gender diversity but also diversity in socio-economic backgrounds, ages, and even philosophical beliefs. After all, the "Singularity of Hope" we envision is one that embraces all facets of the human experience.

Data Auditing and Validation: The Safety Nets

Ensuring that the training data is unbiased is the first step; validating this is the next crucial phase. Techniques like data auditing, where third-party experts rigorously evaluate data sets for biases, should become standard practice. The use of fairness metrics and bias detection algorithms can also serve as computational safeguards. Within our journey toward a hopeful singularity, these are not merely technical exercises but ethical imperatives.

Inclusive Data, Inclusive Futures

In a world heading towards an intimate merger of human intelligence and machine capabilities, the need for fairness and equity in AI is not just a moral imperative but a practical necessity. By committing to inclusive data collection and rigorous validation, we're not just optimizing algorithms; we're optimizing our collective future. This will be the bedrock on which a truly equitable "Singularity of Hope" can be built.

Algorithmic Transparency: The Lighthouse in the AI Fog

As we propel ourselves into a realm where AI and human cognition are entangled, transparency stands as the beacon that guides us through the ethical fog. It's not just about knowing the outcome but understanding the process, the how and the why. If AI is to be our co-pilot in this journey towards an enlightened singularity, we must know how it makes its decisions.

The Open-Source Imperative

Open-sourced algorithms act as the cornerstone for a trust-based AI ecosystem. By allowing multiple eyes to scrutinize the code, we can identify biases, errors, or ethical shortcomings that might otherwise go unnoticed. Open-source is not just a software development methodology; in the context of our book, it's a community-driven approach to ethical governance. It aligns well with the idea that the singularity should not be a closed event but a collective, human-centric evolution.

The Double-Edged Sword: Transparency vs. Proprietary Interests

Yet, we must tread carefully. Openness often clashes with commercial interests. Companies guard their algorithms as proprietary assets, and there's a valid reason for that—these algorithms often represent significant investments of time, intellect, and resources. However, the scales should tip towards transparency when public well-being is at stake. Within the framework of a "Singularity of Hope," proprietary interests should never obscure ethical obligations.

In essence, there may be a middle ground—a 'semi-open-source' model, perhaps, where the core ethical aspects of an algorithm are made transparent while keeping specific proprietary elements confidential.

The Transparency Imperative in the Age of AI

Algorithmic transparency is not just a technical requirement but an ethical one. It's the key to ensuring that AI serves as an augmenting force rather than an obfuscating one. As we move closer to a future where AI and human intelligence coalesce, the lines of code that make up these algorithms will increasingly become the lines that define our ethics, our laws, and our collective mental well-being. Hence, shining a light on these algorithms today will help illuminate our path to a hopeful, ethical, and human-centric singularity tomorrow.

Augmentation with Continuous Monitoring and Feedback Loops: The Living Fabric of Ethical AI

In our journey toward a "Singularity of Hope," where AI serves as an extension of human intellect and creativity, we must recognize that this relationship is not static but dynamic. It evolves, learns, and—hopefully—improves. This is where the beauty of augmentation shines, not as a one-time event but as a continuous process. Just like the human mind benefits from feedback, so should our AI systems.

The Ever-Watchful Eye: Ongoing Evaluation

The very essence of AI is its ability to learn, adapt, and evolve. However, for this evolution to align with human values and ethical considerations, continuous monitoring is not a luxury; it's a necessity. Our AI systems should not just be set up and forgotten; they should be living projects that grow and adapt with us. By continually evaluating the AI's performance and ethical parameters, we can ensure that it stays aligned with human interests, thereby mitigating biases and other ethical pitfalls.

A Symphony of Voices: Community and Stakeholder Feedback

If the singularity is a collective endeavor, then it is only fitting that its ethical scaffolding is also collectively built. Community and stakeholder feedback serve as the 'nervous system' of this ethical construct. By actively seeking input from those who are affected by AI applications, we can tune the system to better reflect diverse needs and ethical standards. In essence, the AI becomes a 'commons,' governed by the collective wisdom of the community it serves.

Augmentation as a Feedback Mechanism

In the book's central theme of augmentation, the relationship between humans and AI is defined as symbiotic. This symbiosis can extend to the realm of ethics as well. Through feedback loops that are built into the augmentation architecture, AI systems can regularly update their ethical frameworks based on real-world outcomes and human input. This allows for a form of 'ethical calibration,' ensuring that the AI system continues to serve as an ethical, augmenting force in society.

Augmentation and Ethics as Living, Breathing Entities

By employing continuous monitoring and feedback loops, and by embracing the ethos of augmentation, we create an ethical AI system that's not frozen in time but is a living, breathing entity. Like a well-tended garden, it requires regular care, evaluation, and the occasional pruning of biases. It is through this ongoing, collective endeavor that we not only prepare for the singularity but ensure that it embodies a future of hope, ethical integrity, and mental well-being for all.

Accountability and Remedial Measures: The Pillars of Ethical Singularity

As we navigate through the complexities of AI and its influence on our lives, accountability and remedial measures serve as the cornerstone of our ethical structure. Just as we stress the need for augmentation and continuous learning, there should be mechanisms to hold AI and its human counterparts accountable. Let's delve into how we can build this accountability into the very fabric of our AI systems.

Establishing AI Ethics Committees: The Socratic Guardians

Drawing inspiration from the ancient Greek concept of a 'philosopher-king,' AI Ethics Committees serve as modern-day guardians of ethical reasoning and societal values in the tech realm. These committees, comprised of experts from various disciplines, stand as a bulwark against unethical practices and biases.

Internal and External Oversight Bodies

Internal oversight ensures that ethical considerations are integrated into the development and deployment phases of AI projects. However, to prevent an echo chamber of ideas, external oversight bodies—comprising independent experts—should also be involved. They can offer unbiased evaluations, adding another layer of ethical scrutiny.

Best Practices: Learning from the Forerunners

Some leading tech companies and research institutions have already set commendable examples. For instance, Google's AI Ethics board and MIT's Media Lab both offer frameworks that prioritize ethical considerations without stifling innovation. These can serve as templates for other organizations aiming for ethical singularity.

Laws and regulations serve as the ethical 'source code,' guiding the behavior of AI systems and their human developers. However, we currently exist in a legal gray area when it comes to AI.

As we approach the singularity, there will be an increasing need for laws that can penalize or prevent biased AI applications. Legislation could range from fines for deploying biased algorithms to mandatory ethical audits for AI systems.

While some existing laws can be adapted to address AI biases, such as anti-discrimination laws, they are not sufficient. There is a glaring gap in the legal framework that specifically addresses the unique challenges posed by AI. Policymakers must act swiftly to fill these gaps, especially as we speed toward a future where AI's societal impact will only magnify.

The Ethical Blueprint for AI

Accountability and remedial measures are not mere afterthoughts; they are integral to our journey toward a hopeful singularity. By establishing robust oversight mechanisms and fortifying our legal systems, we can ensure that our AI systems reflect the ethical and social values we hold dear. It's a collective endeavor, one that demands the active participation of all stakeholders. In this way, we elevate not just the technology but the very essence of who we are and who we aspire to be.

Toward a Fairer AI Future: The Tapestry of Augmentation and Ethical Continuity

As we stand on the precipice of a world powered by AI-augmented intelligence, it becomes imperative to weave the threads of fairness, accountability, and ethics into the fabric of this new world. We're not merely creating machines; we are augmenting the very essence of human existence. And in doing so, we have a profound responsibility to ensure that this augmented world serves us all equally, justly, and ethically.

The Imperative of Proactive Measures

Inaction or passive observation is not an option. We have to be proactive in shaping the ethical landscape of AI. By using augmentation as a tool, we can establish continuous monitoring and feedback loops that not only refine the AI's decision-making but also realign it with our evolving moral and ethical norms. This proactive stance allows us to anticipate ethical dilemmas before they escalate into full-blown crises, making our journey toward the singularity a hopeful and inclusive one.

A Clarion Call for Interdisciplinary Collaboration

The complex ethical issues surrounding AI cannot be resolved by technologists alone. We need philosophers, sociologists, ethicists, and even artists to contribute to this discourse. Interdisciplinary collaboration brings multiple perspectives into the fold, enriching the ethical matrix within which AI operates.

Augmentation and the Feedback Loop: The Symbiotic Path to Fairness

Augmentation isn't just about enhancing our physical or cognitive abilities; it's about enhancing our moral and ethical understanding as well. As our AI systems grow more complex, the augmentation approach calls for a feedback loop—where machine learning meets human ethical reasoning. In this loop, each learns from the other, iteratively refining both the AI's algorithms and our own ethical frameworks.

The Eternal Compass: Our Shared Humanity and Historical Ethics

As we make unprecedented strides in technology, let's not forget the ethical wisdom that history offers. From the moral teachings of ancient civilizations to modern human rights frameworks, we have a rich tapestry of ethical thought to guide us. Our shared humanity and historical ethics should be the compass that navigates us through the murky waters of AI ethics.

Conclusion: A Collective Odyssey

The path toward a fair AI future is not a solitary journey but a collective odyssey, one that requires us to carry along the ethical and moral anchors that have sustained humanity through ages. It's a call to action for each one of us—to not just be passive recipients of AI's offerings but active contributors to its ethical shaping. As we venture forth into this new frontier, let us carry with us the eternal torch of human ethics, illuminating the path for a just, equitable, and hopeful singularity.

Ethical AI as a Mirror of Ethical Humanity: The Reflective Algorithmic Tapestry

As we carry forth our collective odyssey toward a fair AI future, it's critical to pause and reflect on a fundamental principle: Ethical AI is, at its core, a reflection of ethical humanity. The algorithms that power these sophisticated systems are not born in a vacuum; they are sculpted by human hands and guided by human values. In this sense, the ethicality of AI serves as a mirror reflecting the moral and ethical complexion of its creators and users.

The Machine as a Canvas: Painting with Ethical Pigments

When we think of machine learning algorithms, it's easy to become entranced by their technical intricacies. However, behind every line of code, every neural network, and every decision tree lies a set of human-defined parameters. These are the ethical pigments with which we paint the canvas of AI. Just as an artist's choices of colors and strokes reveal their worldview, so too do the ethical assumptions embedded in an AI system reveal our collective moral compass.

Augmentation and Reflective Learning: A Dual Ethical Lens

By following the augmentation approach, we don't just impose our ethical frameworks onto machines; we also allow machines to reflect these frameworks back at us. Through continuous monitoring and feedback loops, we create a dual ethical lens where human and machine ethics co-evolve. This reflective learning enables us to fine-tune not just the machine's sense of fairness but also our own understanding of what fairness means in an increasingly complex society.

The Ethical Feedback Loop: Continuously Refining Our Moral Fabric

In a world where AI is an extension of human capabilities, the ethical feedback loop serves a dual purpose. It not only refines the machine's ethical considerations but also sharpens our own moral clarity. Each ethical dilemma resolved, each bias mitigated, and each equitable outcome achieved adds a new thread to our evolving moral fabric. Over time, this creates a tapestry of ethical thought that is both richly diverse and strikingly cohesive.

Humanity's Ethical Legacy: The Inherited Wisdom in Algorithms

As we navigate the ethical maze of AI, let us not forget the inherited wisdom passed down through generations. Our ethical frameworks are not mere contemporary constructs; they are the culmination of millennia of moral philosophy, social justice movements, and human experience. By embedding this legacy into our AI systems, we honor our ancestors even as we lay the ethical foundations for future generations.

The Fusion of Machine and Human Ethics—A Harmonious Symphony for a New Era

As we draw this chapter to a close, it's crucial to emphasize that the path to ethical AI is a shared journey—one that involves the confluence of machine learning algorithms and human ethical principles. The dynamics between human and machine ethics aren't just a parallel stream; they form an intertwined tapestry that charts the moral landscape of our future.

Augmentation: A Catalyst for Ethical Evolution

Through the lens of augmentation, we can see that AI is not a separate entity that needs to be tamed but rather an extension of human capabilities and, by extension, human values. This means that the ethical frameworks we build for AI are not merely rules imposed on inanimate objects but a reflection of our own ethical and moral commitments. Augmentation serves as a catalyst, accelerating the ethical evolution of both machines and humans.

The Unfolding Ethical Dialogue: A Continuous Journey
The ethics of AI is not a 'one-and-done' affair, solved with a few lines of code or a cursory nod to regulations. It is a continuous dialogue that must adapt to the changing contours of technology, society, and human understanding. Through ongoing evaluation, monitoring, and feedback loops, we can recalibrate our ethical algorithms, ensuring they evolve in step with our collective moral consciousness.

Challenges as Opportunities: Raising the Ethical Bar
It's tempting to see the ethical challenges posed by AI—such as biases, transparency, and accountability—as roadblocks. However, these challenges should be viewed as opportunities for raising the ethical bar. By grappling with these complex issues, we refine our own ethical principles and become better equipped to instill these principles into the machines we create. In the process, we make both ourselves and our AI systems more ethically robust and resilient.

The Eternal Flame of Human Wisdom: Guiding Us Through the AI Era
As we forge ahead, let us be guided by the enduring wisdom of human ethics, honed through millennia worth of philosophical thought, moral dilemmas, and social transformations. These time-tested principles serve as an eternal flame that can guide us through the labyrinth of ethical questions posed by AI. They remind us that while technology may change, the fundamental questions of right and wrong, justice and fairness, remain constant.

The Ultimate Call to Action: Ethical Stewardship in the Age of AI
The task before us is monumental but not insurmountable. It requires ethical stewardship from all stakeholders—developers, policymakers, and

the public. It calls for an unwavering commitment to ensuring that as we advance into an era of unprecedented technological prowess, we do so with the ethical integrity that defines the best of humanity.

In the grand tapestry of human history, the ethical convergence of machine and human signifies not just a technological milestone but a moral one. It's an opportunity to redefine what it means to be ethical in a world where human and machine are becoming increasingly interdependent. Through this harmonious convergence, we are not just coding machines; we are also re-coding ourselves, laying down the ethical tracks for the humanity of tomorrow.

The Convergence of Machine and Human Ethics

Ethical AI is not a distant aspiration but an immediate imperative, and it begins with acknowledging that the machine is a reflection of its creator. By adopting an augmentation-centered approach, enriched by a continuous ethical feedback loop, we can create AI systems that not only meet our present-day ethical standards but also challenge us to elevate those standards. In this harmonious convergence of machine and human ethics, we find a symbiotic relationship that promises to guide us toward a future of unparalleled ethical awareness and collective well-being.

SCAN ME

Made in United States
North Haven, CT
10 November 2026

22091353R00134